THE ITALIAN ECONOMY

CENTRAL ISSUES IN CONTEMPORARY ECONOMIC THEORY AND POLICY

General Editor: **Mario Baldassarri**, *Professor of Economics, University of Rome 'La Sapienza', Italy*

This new series is a joint initiative between Macmillan, St. Martin's Press and SIPI, the publishing company of Confindustria (the Confederation of Italian Industry), based on the book collection MONOGRAFIE RPE published by SIPI and originated from the new editorial programme of one of the oldest Italian journals of economics, the *Rivista di Politica Economica,* founded in 1911. This series is intended to become an arena in which the most topical economic problems are freely debated and confronted with different scientific orientations and/or political theories.

The 1990s clearly represent a transition period in which the world economy will establish new international relationships and in this context, new challenges and new risks will have to be faced within each economic system. Fundamental issues on which economic theory and policy have long based their reasoning over the last two or three decades have to be critically reviewed in order to pursue new frontiers for theoretical development and economic policy implementation. In this sense, this new series aims at being a "place of debate" between professional economists, an updated learning tool for students and a specific reference for a wider readership aiming at understanding economic theory and policy evolution even from a non-specialist point of view.

Published

Mario Baldassarri (*editor*)
INDUSTRIAL POLICY IN ITALY, 1945–90

Mario Baldassarri (*editor*)
KEYNES AND THE ECONOMIC POLICIES OF THE 1980s

Mario Baldassarri (*editor*)
OLIGOPOLY AND DYNAMIC COMPETITION

Mario Baldassarri, John McCallum and Robert Mundell (*editors*)
DEBT, DEFICIT AND ECONOMIC PERFORMANCE

Mario Baldassarri, John McCallum and Robert Mundell (*editors*)
GLOBAL DISEQUILIBRIUM IN THE WORLD ECONOMY

Mario Baldassarri and Robert Mundell (*editors*)
BUILDING THE NEW EUROPE
VOLUME 1: THE SINGLE MARKET AND MONETARY UNIFICATION
VOLUME 2: EASTERN EUROPE'S TRANSITION TO A MARKET ECONOMY

Mario Baldassarri, Luigi Paganetto and Edmund S. Phelps (*editors*)
INTERNATIONAL ECONOMIC INTERDEPENDENCE,
PATTERNS OF TRADE BALANCES AND ECONOMIC POLICY
COORDINATION

Mario Baldassarri, Luigi Paganetto and Edmund S. Phelps (*editors*)
WORLD SAVING, PROSPERITY AND GROWTH

The Italian Economy

Heaven or Hell?

Edited by

Mario Baldassarri
Professor of Economics
University 'La Sapienza', Rome

in association with
Rivista di Politica Economica, SIPI, Rome

St. Martin's Press

© SIPI Servizio Italiano Pubblicazioni Internazionali Srl 1991, 1994

All rights reserved. No reproduction, copy or transmission of this publication may be made without written permission.

No paragraph of this publication may be reproduced, copied or transmitted save with written permission or in accordance with the provisions of the Copyright, Designs and Patents Act 1988, or under the terms of any licence permitting limited copying issued by the Copyright Licensing Agency, 90 Tottenham Court Road, London W1P 9HE.

Any person who does any unauthorised act in relation to this publication may be liable to criminal prosecution and civil claims for damages.

First published in Great Britain 1994 by
THE MACMILLAN PRESS LTD
Houndmills, Basingstoke, Hampshire RG21 2XS
and London
Companies and representatives
throughout the world

A catalogue record for this book is available
from the British Library.

ISBN 0–333–59582–3

Printed in Hong Kong

First published in the United States of America 1994 by
Scholarly and Reference Division,
ST. MARTIN'S PRESS, INC.,
175 Fifth Avenue,
New York, N.Y. 10010

ISBN 0–312–10069–8

Library of Congress Cataloging-in-Publication Data
The Italian economy : heaven or hell? / edited by Mario Baldassarri.
  p. cm. — (Central issues in contemporary economic theory and policy)
Includes index.
ISBN 0–312–10069–8
1. Italy—Economic conditions—1976– 2. Italy—Economic policy.
I. Baldassarri, Mario, 1946–   II. Series.
HC305.I715   1994
330.945'0929—dc20   93–7926
                                                  CIP

THE ITALIAN ECONOMY: HEAVEN OR HELL?

CONTENTS

Introduction: the 80s' Dream, the 90s' Nightmare 7
Mario Baldassarri

Italy: the Real Effects of Inflation and Disinflation 19
Francesco Giavazzi - Luigi Spaventa

Restructuring Processes, Technological Change and Economic Growth 61
Innocenzo Cipolletta - Alberto Heimler

The Role of Monetary and Financial Policies in the Restructuring of Industry... 83
Stefano Micossi - Fabrizio Traù

Disinflation in Italy: an Analysis with the Econometric Model of the Bank of Italy 111
Daniela Gressani - Luigi Guiso - Ignazio Visco

The Government Budget and the Italian Economy During the 1970s and 1980s: Causes of the Debt, Strategy for Recovery and Prospects for Restructuring 157
Mario Baldassarri - M. Gabriella Briotti

Introduction: the 80s' Dream, the 90s' Nightmare

Mario Baldassarri
Università «La Sapienza», Roma

During the past decades, the "Italian case" has often been the subject of attentive analyses that attempt to understand the functioning of an economic system that is, to say the least, "strange". It is strange above all to the eyes of foreign observers who are continually divided between opinions and forecasts of inevitable crises and decline, and then admiration and congratulations for unexpected and surprising successes. But in order to understand the phenomenon of Italy, one has to renounce ex-ante *extreme, one-way evaluations. That is to say, one should not just scan the cover and the contents page, but one should carefully leaf through the pages that document Italy's complex, articulated and often contradictory evolution.*

In reality, Italy's apparent political instability conceals a stability of government, international alliances and internal policy guidelines unparalleled in other leading western democracies and, if anything, is very akin to the Japanese experience.

The enormous gap which separated Italy from the other already industrialised countries at the commencement of its long phase of development immediately after the Second World War, the continuing and evident structural weaknesses that have always been seen over these decades, and the persistent economic-cycle crises, are balanced by a long-term performance which has enabled the country to become, and remain, a member of the group of the world's seven most industrialised countries.

Finally, and more recently, the serious economic, political and institutional crisis of the 1970s was followed, in the 1980s, by greater political stability and an economic, financial, and productive restructuring which engendered new stability and even more substantial prospects for the Italian economy. But here again, perhaps so as not to disappoint those used to the "dual interpretation", the successes of the 1980s were undermined and made precarious by an uncontrolled explosion of public finances which could be seen at the economic-financial level and also in the increasingly obvious gap between the organisation, working, and efficiency of the private productive system and the public-sector players. Although the latter accounts for a considerable share of the economic system, its inefficient public services and inadequate infrastructures strongly limit growth prospects in an international contest of increasing competition between overall systems rather than single industries and/or companies.

By the end of the 1970s, the imbalances of the public finances had already given rise to worried analyses and evaluations (1). Since then, almost yearly, economic policy measures have aimed at "balancing public finances". However, despite these alarms and this continuous attention, it was precisely in these years that the deficit and public debt soared: public debt equalled 100% of GDP in 1989 with a continuous exponential path in the successive years.

Almost invariably, every year saw the ritual of an uncontrolled rise in spending decided (and often concealed) in the first half of the year, followed by a desperate resort to fiscal revenues with "summer decrees" in August, then the drafting of a budget in September, and finally a "Christmas decree" at the end of the year.

In concrete terms, however, the numerous measures adopted have been only partially successful, for while containing increases in the deficit and debt, they led neither to their stabilisation nor even their reduction. Moreover, these measures for relative containment were implemented by cutting back, in effect, on the strategic re-

---

(1) See in the Bibliography of the paper by M. Baldassari- G. Briotti: BERNARDI L. [8]; CAVAZZUTTI F. [11]; COTULA F. - MASERA R.S. - MORCALDO G. [14]; MAJOCCHI A. [18]; SPAVENTA L. [23].

sources needed for those large-scale and modern infrastructures that are essential for the overall efficiency of a country's productive system. In other words, we continued to pay wages and state pensions without readjusting the fixed corporate capital. Moreover, changes in the foreign exchange and monetary policy guidelines, which played an important role in forcing the productive system to restructure, have triggered a vicious circle of interest payments/debt in the national budget.

The favourable international situation and the success of the domestic manufacturing system have to date, thanks to sustained and durable growth rates, prevented the situation from degenerating into a dangerous financial crisis. Economic theory has analysed at length the problem of the sustainability of the public debt (2), which, however, remains an empirical problem and one that concerns the more general credibility of the institutions and their capacity to govern the system. It is not unarguable, however, that a different international situation and growing domestic difficulties could significantly curb the growth potential, and thus give rise to worrying prospects of financial crises.

In this sense the Italy of the 1990s would appear to resemble a robust lame duck: while it can still fly, it can achieve long trips and reach high altitudes; but should it be forced to land, it risks seriously injuring itself and finding it more difficult than the others to take to the air again.

A consistent slowdown or, even worse, an international recession would inevitably risk upsetting the precarious equilibrium of the Italian economy. Hence, while not denying the Italian economy's considerable capacity to react and recover, the domestic and international situations which would appear to lie ahead in the 1990s place precise constraints and, unfortunately, could also herald concrete conditions of risk.

This is why it is appropriate to collect in this monograph a series of essays which enable one to relate the various interpretations given

---

(2) See in the Bibliography of the paper by M. Baldassari-G. Briotti: ALESINA A. [1]; ALESINA A. - PRATI A. - TABELLINI G. [2]; BARRO R.J. [7]; BLANCHARD O. - DORNBUSH R. - BUITER W. [9]; CIVIDINI A. - GALLI G. - MASERA R.S. [13]; SPAVENTA L. [23]; TABELLINI G. [26].

to the industrial restructuring phase to the importance, role, and structural effects of a national budget, especially one that is progressively undermining the country's productive base and financial equilibria. Dangers of a financial crisis and a deindustrialisation process would therefore appear to be the "big risks" against which the Italian economy's concrete prospects for growth over the coming years should be evaluated.

Italy's economic development in the difficult transition from the crises of the 1970s to the productive and financial restructurings of the 1980s has been the subject of profound and keen analysis by several scholars who have studied the various aspects of a phenomenon which is undoubtedly complex.

The productive restructurings and the conditions that engendered them are examined in the first four papers of this collection which opens with an essay by Giavazzi and Spaventa.

The two authors maintain that the roots of the present level of public debt date back to the budget imbalances that originated from the social welfare reforms in the early 1970s. They argue that the restructuring costs of the early 1980s were actually largely met, and that the causes of the present debt have, to a considerable extent, nothing to do with the productive restructurings of the 1980s. The logical sequence of the line of reasoning advanced by Giavazzi and Spaventa can be summarised as follows: in the second half of the 1970s, the fiscal-drag which followed on the heels of the effects of the fiscal reform led to a growth in incomes that in turn made it possible to transfer resources to support the restructuring of the production system.

This essay is followed by an analysis by Cipolletta and Heimler which, while not denying the positive contribution of the economic policy demonstrated by Giavazzi and Spaventa, shows that the quantitative aspect of the measures adopted to support the productive system was very modest, vis-à-vis the dimensions of the investment and restructuring process which largely concerned industry. The impressive process of transformation of the Italian production system cannot, therefore, be attributed to state intervention. One should also recognise the role played by the exchange-rate policy and the consequent monetary policy which, as Micossi and Traù main-

tain, placed very rigid strictures, thus opening the way to restructuring as the only and proper structural solution.

The deflationary process that could be observed in Italy in the first half of the 1980s (with inflation falling from 22% in November 1980 to 4.2% at the end of 1986) is the subject of an articulated analysis in the essay by Gressani, Guiso and Visco. This paper uses the Banca d'Italia's quarterly model in an attempt to measure the contribution to deflation of what would appear to be the four most important reasons, namely: 1) the forces pushing and braking on prices 'imported' into Italy from abroad mainly as a result of the energy prices and the dollar exchange-rate trends; 2) the monetary and exchange-rate policies adopted by Italy during the years in question; 3) the policy of "predetermination" and "planned inflation ceilings"; and 4) the impulses deriving from public spending, tariff policies and indirect taxation.

The author's conclusions can be summarised as follows:

a) *foreign price trends had a considerable influence on the Italian inflation rate given the economy's level of opening and the presence of widespread indexation mechanisms;*

b) *the recovery process was, nevertheless, the result of a continuous fall of the domestic component rather than a diminution of the foreign impulses, with the exception in 1986 when the effects of the steep fall in oil prices and the depreciation of the dollar were especially significant;*

c) *monetary policy, including both exchange-rate policies and control of interest rates, played a decisive role in reducing the inflation rate. In particular, the adoption of an accomodating exchange-rate stance would have intensified the inflationary impact of the second oil shock and led to a pattern of inflation far worse than that seen, with only modest gains in income. Moreover, although pegging exchange rates to the nominal rates in force prior to the second oil shock, while apparently possible, would have had, on the one hand, considerable deflationary effects, it would have led to considerable losses of income and a worsening of the external accounts that would have probably made it impossible to maintain such a policy;*

d) *the policy of predetermination and, more generally, the announcement of planned inflation rates by the government would appear to have had limited effects as regards deflation. It is nonetheless probable that it contributed to avoiding the rise of autonomous impulses that could have hampered the recovery process and, more importantly, would have increased the costs of the monetary and exchange-rate policies, while also making their implementation more difficult».*

It is our opinion that these various interpretations do not exclude one another, but taken together comprise a range of analyses and interpretations that enable a better and more detailed knowledge of the reality of the situation.

The analysis in the paper by Baldassari and Briotti which concludes the collection, tends to emphasise the difficulty, and perhaps the impossibility, of precisely dating and identifying the origins of Italy's public debt. The authors also assess the encumbrance and constraints which the state finances place on today's Italian economy.

The paper emphasises an evident fact, namely that the range of objectives pursued by the public sector far exceeded the resources available. Furthermore, pursuing these various objectives, even with a significant increase in fiscal revenue, led to a growing deficit, not only overall but also in the current account.

This mode of pursuing objectives exceeding the resources available finally had the perverse effect of reducing the resources which were potentially available.

In reality, the first phase of the social reforms of the 1970s was accompanied by a phase of state entrepreneurship both directly, in the state-owned company sector, and indirectly, and perhaps necessarily, in the private sector.

The problem is therefore not so much that of attributing one or another party with the responsibilities for the present situation of public finances or the merits for the production restructuring process which took place.

In reality, an imbalance between objectives and resources means that at the end of the day someone has to foot the bill. The accumulation of debt means postponing the settling of the "general account" to future generations. Moreover, as the paper seeks to *demonstrate, this process grows in an exponential manner and, perversely, progressively reduces potential resources. Postponing matters to what ten or fifteen years ago was the "long term" has therefore now become an everyday phenomenon.*

Hence it is asked who will eventually have to foot the bill and to what extent today's public debt stock reflects the absence of such a payee, while faced with a payee who only apparently, and with perverse effects, pays the outstanding accounts.

The public debt is today's "collection tin", that is to say, the parameter which reveals not only the restructuring which is incomplete, but more importantly that which has still to be implemented.

It is certainly not easy, particularly for politicians, to realise that the old mechanism of deficit-spending can no longer work and hence draw the necessary drastic conclusions in terms of balancing the deficit and public debt according to a schedule can now only be rapid, in view of the imminent "European" deadlines.

In many ways, what was in past decades called the "rising star of Italy" can, at least ex-post, *be understood and perhaps even rationalised. Deficit public spending allowed the distribution of resources, the gathering of consensus, and the easing of social conflicts without "simultaneously and apparently" demanding sacrifices from anyone. The long surge of growth and international integration, to which far-seeing political choices made in the 1950s and 1960s had coupled the Italian economy, allowed a relative balancing of the national accounts thanks to a strong growth rate and hence an automatic increase in revenues. Inflationary differentials could be corrected with exchange-rate measures, and business-cycle crises were expeditiously dealt with and resolved. In fact, when seen from a long-term viewpoint, the crisis of 1963-64, which at the time appeared very serious and dramatic, was nothing more than a brief pause for reflection that was dealt with in time and rapidly resolved. It is true that the favourable international conditions also then helped the Italian economy to resume its path of balanced growth, but it is equally true that in that now long-distant experience, the government and the Banca d'Italia showed themselves capable of reacting promptly with measures which were appropriate both qualitatively and quantitatively.*

However, that "toy" and perhaps also that "model" began to crumble at the start of the 1970s: radical changes took place on the international scene, while on the domestic front state finances saw the perverse deficit-spending mechanisms resorted to with an ever increasingly frequency.

Italy's present-day economic policy should therefore rediscover, as the corporate jargon has it, a strategic "focus". In fact, one of the most serious aspects of the present situation (which creates the requirements for the two big risks that the Italian economy runs in the 1990s: a financial crisis triggered by the public debt, and the consequent deindustrialisation of the system) is not only the game of tag seemingly played by the figures on the state balance-sheet and the periodic failure to achieve the containment targets, but rather the fact that Italy appears to be a country "without a strategy". And unfortunately the crisis and the Gulf War risk inducing a "soporific" effect. The international recession feared during 1990 had at least ensured the ringing of important alarm bells. Now, thanks to the effects of the post-war period, the recession, in 1991, would appear to be resolved and the risk is one of "after the storm ..." . But a country without a strategy is like a person exposed to carbon monoxide who apparently has a "painless" death and, if nobody turns the gas off, is discovered by the concierge the morning after, too late for anything to be done.

The 1992 events seem to be the "concierge's alarm": the lire has been sharply devaluated and the possibilities of Italy joining the European monetary union appear increasingly distant and difficult.

With regard to real and industrial strategies, the economic and financial wrongfooting created by the public finances leads to, on the one hand, a structure of large-scale networks, both old and new communications, transport, administration, etc.) that are completely inadequate vis-à-vis the competitiveness of the system which characterises the 1990s. On the other hand, it prevents the creation of the driving financial forces (self-financing, bourse, credit) which are able to cope with the new, and in many sectors gigantic, dimensions being consolidated in the new oligopolistic international set-up. One should not leave out the costs and time required to enable Italy's banking system to cope with the technological changes in processes and product. These lengthen the "return" times of the necessary investments in fixed capital, organisation and human capital. In the short- and medium-term this leads to higher trading costs (3).

In these conditions, what and who can make, in the broadest and most comprehensive meaning of the word, an industrial policy? Many industries' strategy revolves around a new accelerated process of concentration that has resulted, for example in the consumables sector, with 60-70% of the world market being held by no more than 10 or even 5 companies. The forces driving this process would appear to differ from sector to sector: high production costs, in the case of the aeronautical sector; production and above all distribution economies of scale for white goods, high definition in the television sector, production integration in the engineering, automotive, automation and control sectors; and high-speed in the railway sector. Differing reasons which nevertheless have the same result: a process of concentration which places the power of defining strategies and development in the hands of a few, while perhaps simplifying decisions concerning the apportionment of production between the various countries of the world to the easy and changeable conditions of comparative advantage in production costs. It vis-à-vis these processes that the Italian system reveals its considerable strategic limits and risks a deindustrialisation which, yet again, has "soporific" features: we are not talking about a sudden industrial slump but rather a gradual loss of contact and role with (and within) the new international oligopolies. It is by no means impossible, in fact it is very probable, that some Italian industrial players will reveal themselves capable of staying in the world's oligopolistic club. The problem, however, is that if these players do their sums "by consolidating between themselves", the Italian system as a whole risks appearing as nothing but a "production field" where the strategies are decided "somewhere else".

Hence we are not talking about tactical corrections to be made to the overall situation of public finances. To avoid the "big risks",

---

(3) See Lo Cascio M. - Malavasi R.: *Innovazione e progresso tecnico nel sistema creditizio italiano*, Cedam, Padova, 1989.

we are nowadays faced with the "evident need for a combined corrective measure which, following the example of any company faced with a similar experience, should be based on adjustments to the "revenue account" and operations concerning its "assets and liabilities".

The first aspect entails realigning social spending to current revenue, containing the former's growth, while, with regard to the second aspect, selling part of the state's assets. We are certainly not talking about "selling off" anysthing, but simply recognising that the state assets are in reality not "public" since the black-ink items in the statement of assets and liabilities are more than offset by a considerable public debt. Hence the state's activities belong in reality to those who hold Bot and Cct treasury bonds. Furthermore, the state is in the perverse situation of borrowing at interest rates that are considerably higher than the rates of return it obtains on the black-ink items of its own statement of assets and liabilities, while also assuming the corporate risk and guaranteeing the lenders a secure return. However this is all on a purely corporate plane. On a macroeconomic plane, similar conclusions are obtained where the real interest rate is greater than the economy's rate of growth.

It should also be noted that the sale of part of the assets should be used to reduce the liabilities, and not be confused with the revenue account (the positive effect on the public deficit *could derive from the consequent reduction of interest rates); this sale should take place via the market and hence can only concern activities that have a market value, following a meticulous verification of whether or not they are strategic for the achievement of the state's objectives, and carefully evaluating the risk of excessive oligopolistic or monopolistic concentrations in the private sector.*

One of the collection's fundamental arguments is that there is a disparity between the range of the state's objectives and the resources available. There is, however, another element not to be overlooked that is not dealt with here, namely that of efficiency and effectiveness. The three objectives, social state, entrepreneurial state and a state which invests in infrastructures, can be compatible with each other even with equal resources if their efficiency is improved. This also requires the introduction of private-sector and market parameters, bearing foremost in mind that logic would demand that priority be given to spending on infrastructures which, by opening better growth prospects to the production system as a whole, would create the resources needed to sustain the social state. The entrepreneurial state follows, but with its feet securely "in the market".

Italy: the Real Effects of Inflation and Disinflation

Francesco Giavazzi - **Luigi Spaventa**
Università di Bologna   Università «La Sapienza», Roma

1. - Introduction (*)

Over the past 15 years, the Italian economy has experienced rather uncommon developments, by comparison with the rest of Europe. These developments have been described alternatively in terms of near-collapses, miracles, or impending bankruptcy. Such dramatic characterizations are of little help in understanding what happened. Yet, they somehow reflect a history of large and often unexpected changes and a sequence of policy turn-arounds. From a distance, the Italian experience still raises some issues of general interest regarding the means and the timing of an adjustment process.

(*) The authors are grateful to Elhanan Helpman, Rainer Masera, Charles Wyplosz and Fabrizio Barca for discussions, to Stefano Fantacone for research assistance, and to the Istituto Centrale di Statistica for helping with the data.

Italy's national accounts recently underwent a very thorough revision which affects not only the levels but also the time profile of the relevant variables. The revision incorporates in the accounts previously unrecorded activities, as they emerge from census data and from a more comprehensive definition of employment. Employment is now defined in terms of 'units of labour', which, unlike previous data, include part-time and irregular, work, standardized for the time actually worked, while excluding workers on subsidized lay-off (*Cassa integrazione*). Whenever possible, we present our statistical material and our computations with reference both to the old (o.n.a. in the tables) and to the new (n.n.a.) data, as the latter are not yet published in international sources. Notes to the tables warn when only old data are used because the new ones are not yet available.

This paper was published in *Economic Policy*, n. 4, 1989 and in Italian in *Rivista di politica economica*, Roma, Sipi, July-August, 1989.

In Italy, return to profitability was pursued before macroeconomic stabilization, instead of resulting from it, as happened elsewhere. It is indeed our contention that this evolution is sometimes commendable: adjustment in Italy was quicker and less costly, in terms of output, investment and employment.

The supply shocks of the early 1970s caused severe domestic and external imbalances, which were perhaps even more pronounced in Italy than in other European countries. The policy response to the shocks was a peculiar blend of supply-side measures, inflation and exchange-rate depreciation (1). The recovery of output was stronger and the turn-around in profitability happened earlier than in other 'healthier' countries, fostering a record investment boom. In the early 1980s, as inflation was accelerating, European Monetary System discipline was belatedly accepted and was used to enforce stabilization. One might presume that the cost of disinflation in terms of output would offset the gains of the earlier boom, so that the inflationary spell would leave output unaffected in the long run. This, however, was not the case. There was indeed a recession, but it was short and relatively mild. Since 1983 output and investment have grown at a faster pace than in other major European countries, while the inflation differential with respect to Germany has narrowed considerably (Table 1). In terms of output, Italy has clearly outperformed other European economies in the period 1976 to date.

Our emphasis on the role of policies indicates that the popular explanation of Italy's growth in terms of the strength of its underground economy is, in our view, wholly unsatisfactory. The possibility of evading rules imposed by the tax authorities or the Labour Office may be relevant in other contexts, but can hardly account for the reaction of the economy to wage and oil shocks or to the imposition of EMS discipline.

On the whole, we find that Italian policies have gone beyond simply postponing the output cost of adjustment to the supply shocks; it is not simply a matter of paying later what other countries had paid

---

(1) The role of policies in this period is stressed by GRAZIANI-MELONI [20], CER-IRS [11], BARCA [4], CIPOLLETTA [13], ANDREATTA-D'ADDA [1] provide an opposite, and critical appraisal of policies in this period.

Advise: the numbers in square brackets refer to the Bibliography in the appendix.

TABLE 1

INDICATORS OF PERFORMANCE
(average annual growth rates)

	1977-1980	1981-1982	1983-1987	1977-1987
Gross domestic product:				
France	2.8	1.5	1.6	2.0
Germany	2.9	0.4	2.2	2.1
UK	1.1	− 0.1	3.3	1.9
Italy (o.n.a.)	3.3	− 0.2	n.a	n.a
(n.n.a.)	4.3	0.6	2.6	2.9
Gross fixed capital formation: total				
France	1.7	− 1.6	0.3	0.5
Germany	4.6	− 5.0	1.8	1.5
UK	− 0.4	− 2.3	4.4	1.4
Italy (o.n.a.)	3.6	− 2.3	n.a	n.a
(n.n.a.)	4.2	− 3.7	2.8	2.1
machinery and equipment:				
France	2.8	− 0.8	1.7	1.7
Germany	6.9	− 5.5	4.5	3.4
UK	3.5	− 4.8	6.2	3.1
Italy (o.n.a.)	5.7	− 3.4	—	—
(n.n.a.)	9.1	− 4.4	5.7	5.0
Private consumption deflation (difference from Germany				
France	6.2	6.8	4.1	
UK	9.4	4.6	2.9	
Italy (n.n.a.)	12.3	12.2	7.7	

Sources: *European Economy*; for investment: OECD: *Quarterly National Income Accounts*, n. 3, 1988.

earlier. Rather, the sequence of policies has affected, and in our view improved, the outcome. A comparison with Britain, which experienced a similar situation but different policies, is instructive and can be conveniently summarized by Graphs 1 and 2. Italy chose to boost firms' profitability first, tolerating inflation — even letting it increase considerably. As a result investment soared, creating the conditions for deep structural adjustment in industry. Instead, Britain which faced similar inflation pressure, elected to deal with it first, which resulted in sagging profitability and investment. The structural adjustments required by the supply shocks thus occurred in the context of major hardships, as indicated by the growth performance shown in

GROSS PROFIT SHARES (*)

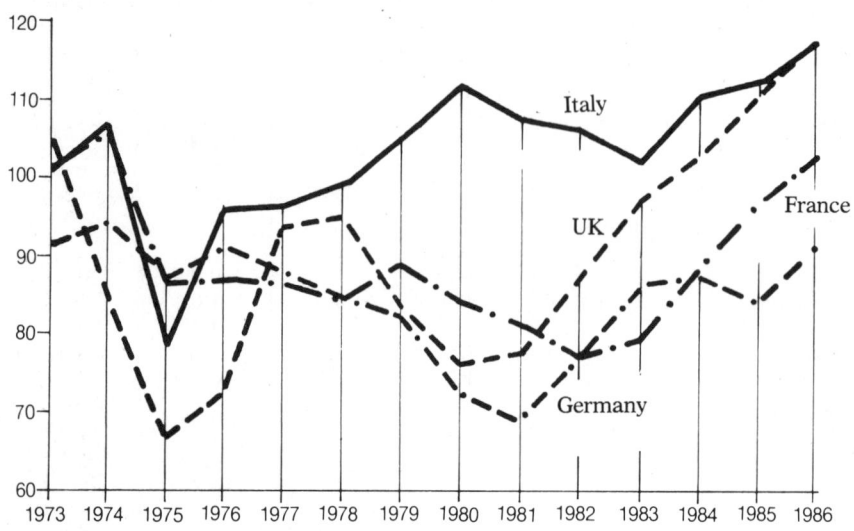

(*) Shares in value added at factor costs adjusted for imputed imcome of the self-employed.
Source: See Table 4.

Table 1. With comparable initial conditions and final outcomes (in terms of output and inflation), Italy has over the whole period created more income and presumably enjoyed higher welfare.

This leaves an important issue open; the question remains whether this favourable outcome was obtained at the expense of high deficits and mounting debt. If so, the public debt accumulated as a counterpart of the relatively good performance of the economy would represent a bill yet to be settled and the story would look much less successful. While fully recognizing the relevance of the debt problem, we shall however argue that its origins cannot be imputed to the policies we discuss in the paper: the debt issue has a history of its own, quite unrelated to the policies we examine.

In the next section we consider the shocks of the first half of the 1970s, and the problems facing policy makers at the time. We show

INFLATION DIFFERENTIALS WITH GERMANY
(consumer prices)

GRAPH 2

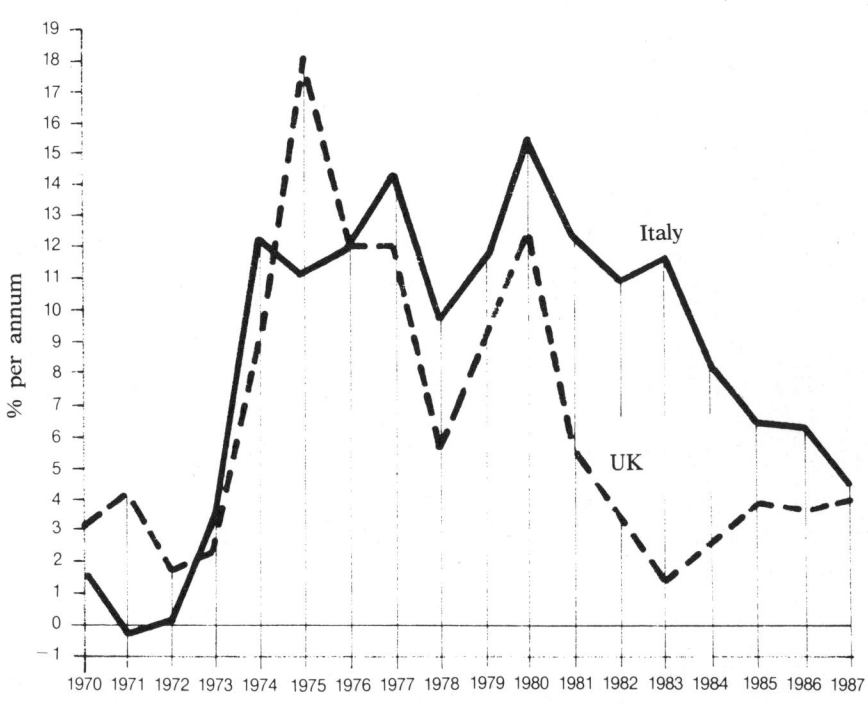

Source: OECD: *Economic Outlook*.

how a mix of inflation, currency depreciation and supply-side measures can provide a second-best solution in a regime where real wages are rigid downwards. Section 3 deals with the return to the straight and narrow path of EMS. This is, in a way, a more conventional story but one which accounts for the lower output cost of Italian disinflation. In section 4, we explicity examine the debt issue, and we show that the problem of budget deficits, which translates into high current indebtedness, has roots in an earlier period. In the last section we draw some conclusions.

2. - The Shocks and the Policy Response

2.1 *The Shocks and the Problems*

In the early 1970s supply shocks affected all European economies. Nominal wages and unit labour costs accelerated in all major countries, and more so in Italy and the UK. Fast output growth, while offsetting to some extent the consequences of rising wages, was accompanied by a hike in raw materials prices. The short-run effects on profit shares varied across countries, depending on the extent to which an inflationary response was tolerated. Between 1973 and 1974 the oil shock affected Italy more than other countries due to greater dependence on imported oil. To illustrate, in 1974 oil accounted for 56.5% of energy consumption in Italian industry as compared to 51.2% in France, 45.8% in the UK and 36.9% in Germany. As a result, total requirements of imported oil represented 4.6% of final demand in Italy, 4.6% in the UK, 3.3% in France and 2.8% in Germany (data from Silvani [29]). The additional "oil bill" imposed by the oil shock (higher cost of imported oil minus additional exports to oil producers) was in fact greater in Italy than anywhere else, with the exception of Japan.

The analysis of supply shocks is by now familiar (see for example, Bruno and Sachs [9]). To avert negative effects on output, downward flexibility of real incomes, and in particular of real wages, is required; indeed, wage indexation raises inflation, while higher demand does not restore the previous level of output unless real wages decline. As a result, stagnation or recession can be avoided neither by expansionary policies, nor by exchange-rate depreciation, which feeds right back into prices because of indexation. The situation improves with time only as the pressure of unemployment removes the rigidities in the labour market and real wages decline. Rapid substitution of capital for labour may restore the output level in spite of rigid real wages, but at the cost of higher unemployment.

Nowhere in Europe were real wages flexible in this period. In Italy the rise in nominal labour costs was, however, greater than elsewhere and accelerated precisely in connection with the oil shock.

Strengthening of formal indexation was agreed upon in 1975;

over and above this, high wage increases were requested, and granted, in a situation of widespread labour unrest; rigidities in the use of labour also became worse. Tables 2 and 3 show the extent of the problem; in Table 2, changes in the price of manufacturing output are disaggregated into three sources, namely labour costs, profit margins and the terms of trade between materials and output prices (assumed to be exogenous). In 1974 the terms of trade factor was responsible for almost four percentage points of the price rise: still, even in that year, the contribution of labour costs amounted to almost 16 points. In spite of this, accommodating monetary policy (in a situation of excess demand) allowed a rise of profit margins and of the profit share. In the following two years, labour costs rose even faster and were responsible for almost the entire rise of output prices, while the profit share fell abruptly.

TABLE 2

CONTRIBUTIONS TO AVERAGE
ANNUAL CHANGE OF OUTPUT PRICES IN MANUFACTURING (*)

	1974	1975-1976	1977-1979	1980
Labour costs (a)	15.8	16.7	10.3	9.8
Profit margins (a)............	8.9	2.4	4.7	7.3
Terms of trade	3.8	− 1.7	− 0.1	0.7
Residual	0.9	− 0.2	− 0.1	0.2
Rate of change of output prices	29.4	17.2	14.8	18.0
Memorandum items:				
Rate of change of input prices	37.2	14.9	14.6	21.2
Shares on value of output				
labour	− 2.8	3.4	− 2.0	− 21.2
profits	0.9	− 1.6	2.2	1.2
external inputs	1.9	− 1.8	− 0.2	0.4

(*) o.n.a.: value added at factor cost; (a) adjusted for self-employed and for subsidized lay-offs. Let q be manufacturing output and p its price. Then $pq \equiv W + \pi + M$, where W are gross wages, π gross profits and M the value of inputs external to the sector. Lower-case letters denote ratios to output, with α and β respectively the labour and the external inputs per unit of output, and p_M the price of external inputs, then: $p = \alpha w + \pi + \beta p_M$. If the terms of trade, p_M/p, are exogenously and $\gamma = (1 - \beta p_M/p)^{-1}$, the reciprocal of the share of value added on the value of output: $p = \gamma(\alpha w - \pi)$, and: $\dot{p}/p = \gamma(\alpha w/p)\dot{w}/w + \gamma(\pi/p)\dot{\pi}/\pi + \dot{\gamma}/\gamma$.

Sources: COMMISSION OF THE EUROPEAN COMMUNITIES (Directorate General for Economic and Financial Affairs): *Indicators of Profitability, Capital, Labour and Output*, mimeo, 1988. Number of workers on subsidized layoffs: authors' estimates, 'Ratio of number of subsidized hours to yearly contractual hours', data from BANCA D'ITALIA: *Annual Report*, various issues, and BARCA [4].

TABLE 3

LABOUR COSTS IN MANUFACTURING
(annual compound rates of change, %)

	1974	1975-1977	1978-1980	1981-1985
Nominal cost of labour per employee:				
France	12.6	17.9	14.0	10.4
Germany	11.6	9.2	6.2	5.1
UK	15.6	20.5	15.6	10.3
Italy (o.n.a.) *(a)*	25.8	22.5	16.9	16.3
(n.n.a.) *(b)*	26.1	21.9	17.6	16.3
Real value added per person employed *(c)*:				
France	2.6	4.3	3.4	3.0
Germany	2.7	4.8	0.5	3.4
UK	1.8	1.4	− 1.3	5.8
Italy (o.n.a.) *(a)*	4.0	1.5	5.1	3.4
(n.n.a.)	3.6	3.0	5.8	4.3
Nominal cost of labour per unit of real value added:				
France	9.7	13.0	10.2	7.2
Germany	8.7	4.2	5.7	1.6
UK	13.6	18.8	17.1	4.2
Italy (o.n.a.) *(a)*	20.9	20.7	11.2	12.5
(n.n.a.)	21.7	18.4	11.1	11.5
Real labour cost per unit of real value added *(d)*:				
France	− 5.3	3.0	0.1	− 1.7
Germany	0.4	0.5	2.2	− 1.6
UK	10.1	− 1.4	1.7	− 2.2
Italy (o.n.a.) *(a)*	− 6.2	2.4	− 3.3	0.7
(n.n.a.)	− 3.1	1.8	− 1.6	− 0.1

(a) the o.n.a. data on employees in employment and on total employment are corrected for dependent workers on subsidized layoffs;

(b) units of dependent labour;

(c) employees plus self-employed. For Italy (n.n.a.) total labour units, value added at market prices;

(d) deflated value added at market prices.

Sources: COMMISSION OF THE EUROPEAN COMMUNITIES (Directorate General for Economic and Financial Affairs): *Indicators of Profitability, Capital, Labour and Output.*

Table 3 provides comparisons with other European countries (in terms of manufacturing value added rather than output). In 1974, nominal labour costs rose in Italy much faster than anywhere else, but high inflation allowed a temporary decline in real unit labour costs. In the following two years, the nominal cost of labour per unit of real value added rose faster also, because of a disappointing performance of productivity: a higher rate of inflation (17-18%) was not sufficient to prevent a rise in real labour costs and a sharp decline in the profit share (Table 4).

TABLE 4

CHANGES OF ADJUSTED GROSS PROFIT SHARES
IN MANUFACTURING (%) (*)

	1973-1974	1974-1977	1977-1980	1980-1985
France	+1.5	−6.3	−0.8	+4.2
Germany	−0.2	−0.8	−4.4	+3.3
UK	−5.0	+2.3	−4.3	+8.4
Italy (o.n.a.)	+2.6	−5.2	+7.9	−2.5
(n.n.a.)	+1.8	−3.1	+4.6	+0.1

(*) Shares of value added at factor cost adjusted for imputed income of the self-employed.
Source: See Table 3.

Domestic inflation was accompanied by huge external deficits. Between 1974 and 1976, Italy had to have recourse to the EEC support arrangement and to an EEC recycling loan, to an IMF stand-by arrangement and to the IMF oil facility, to a swap of gold for currency with the Bundesbank and to a swap arrangement with the Federal Reserve. There were repeated currency crises and the exchange market had to be closed for over a month at the beginning of 1976. In short, up to 1976-1977 Italy displayed the symptoms predicted by theory for an indexed economy hit by a supply shock. Few at the time would dispute the conclusions of the 1977 OECD *Survey of Italy* (OECD [26]): «At the beginning of 1977, the situation of the Italian economy remains precarious. There is general agreement in the country on the need to implement an effective stabilization program ... [But] no lasting agreement has been reached ... on the choice of a method to stop the incomes-prices spiral ... The sacrifices implied by the indispensable austery measures risk to be ineffective if a pace of inflation of the order of

20% is going to persist for long. Radical stabilization measures are inevitable to avoid that domestic inflation and currency depreciation reinforce each other and that the adjustment of the current balance ... is only transitory. Considering the nature of the disequilibria ... they will not however be sufficient to bring the economy back onto a path of balanced growth».

The IMF staff's analysis prepared for the second stand-by arrangement in the Spring of 1977 criticized deficient policies; according to that analysis: «another year of low growth ... is inevitable if Italy is to break free from constrictions ... that presently hamper economic policies»; in addition, the IMF expressed «the belief that the nature of the disequilibrium ... is such that correction in a short period of time is not possible» (Spaventa [30]). In the OECD's view, unless the indexation mechanism was profoundly changed, policies of demand management could achieve stabilization only at the expense of employment and investment, while exchange rate depreciation would cause an increase in inflation (2).

These views should be compared with the picture of the period 1977-1980 which emerges from Graph 1 and Tables 1 to 3. There was a strong recovery of output, led by exports and strengthened by an investment boom. Output grew faster than elsewhere; more importantly, the turn-around in profitability and in the wage gap (however measured) occurred in 1978 — much earlier than in other countries (Bruno [8]) and United Nations [31]). Real unit labour costs fell substantially in Italy, but rose elsewhere — in low inflation Germany as well as in high inflation Britain. Inflation in Italy declined somewhat but remained far higher than elsewhere; gross nominal wages kept rising at over 20% and nominal exchange rate depreciation continued. The recipes recommended by international organizations were followed to a very limited extent: one many thus wonder why their predictions did not come true. To what extent were inflation and depreciation instrumental in reaching a relatively favourable and largely unexpected outcome? We shall now try to answer these questions.

---

(2) OECD [26], MODIGLIANI-PADOA SCHIOPPA [25] provide a formal analysis.

2.2 Subsidies, Taxes and Depreciation

We intend to show how an unconventional mixture of subsidies and exchange-rate depreciation can be used in an indexed economy hit by a supply shock to yield the following results: a favourable effect on the supply side, by restoring profitability; a favourable effect on the demand side, by inducing a real depreciation and thereby stimulating exports; and additional revenues to finance the subsidies. We first tell the story in general terms (formalized in the Appendix) and then show how the story applies to the Italian case.

2.2.1 A Simple Parable

We imagine a country which produces manufactured goods using only labour and imported "oil". Initially, this country only trades with the oil exporter, exchanging its output for oil. There is now a sudden rise in the *real* oil price. With wages fully indexed to the price of the home produced goods, the shock affects only the profit margins, leaving real wages unchanged. As the size of the fall in real margins is independent of the price level, inflation is by itself wholly ineffective. The fall in real margins will probably depress investment and cause a slump.

At the macro level there is thus nothing government can do. The government could, however, decide to enact a "supply-friendly" policy, amounting to subsidizing a recovery in margins. Typically, this can be done by reducing the wedge between take-home pay and the cost of labour for the enterprise, either by lowering taxes on employees — *is* they accept a corresponding reduction of their gross wage — or, more directly and more safely, by reducing social security contributions, which is equivalent to granting a subsidy proportional to the wage. The pitfall of this latter solution is of course an increase in the budget deficit. How can the government avoid the budgetary cost without having to ask Parliament to raise taxes to subsidize profits (a difficult proposition in a pre-Thatcherite era)?

The answer may come from personal income taxation. With a proportional income tax nothing can be done: if the increase in nominal incomes is in line with the rate of inflation, real revenues remain the same and higher nominal revenues finance the nominal increase of unchanged real expenditures. If instead the income tax is progressive and tax brackets are fixed in nominal terms, tx rates increase with a rise in income, whether nominal or real: this introduces another variable in the system, which like profit margins is not indexed. This is important, since *real* revenues now depend on the price level. As long as *only the gross* wage is indexed or, more generally, as long as wage earners do not pass on into higher wages an increase in taxation due to inflation, a price rise provides the government with an increase in real revenues, which can be used to subsidize profits (more elegantly, to cut social security contributions). In principle, supposing only for a moment that it can fully control the rate of price rise, the governement could set the latter at a level such that the increase in real revenues is enough to finance the subsidy needed to offset the fall in real margins caused by the shock with full gross wage indexation. Note, again in principle, that to ensure a *permanent* rise in real revenues through progressive taxation, what is required is not an increase in the *rate* of inflation, but a *one-shot* price rise, which is sufficient to allow a *permanent* rise of real tax rates.

If the home country also trades with another industrial country whose prices are supposed for simplicity (but without loss of generality) to be fixed, home wages are indexed on a basket of home produced goods and of imports from the other industrial country and their change will depend on domestic inflation and on the change in the nominal exchange rate. Hence, wages do not fully reflect domestic inflation and, for a given nominal exchange rate, an increase in domestic prices lowers the product wage at a constant consumption wage. An increase in the price level can thus be used to raise real margins: the cost is a real exchange rate appreciation, and consequently a loss of competitiveness and a fall in export demand. Here again, subsidies can be used to restore real margins to their pre-shock level; to pay for the subsidies, the increase in the price level must, however, be accompanied by a depreciation of the nominal exchange rate, so as to keep the real exchange rate constant. In the same way as there is a level of subsidies that allows real margins to remain constant with a constant real exchange rate, higher subsidies should achieve the same recovery of margins *and* a real depreciation. With real depreciation, subsidies will of course need to be higher; to finance them, nominal depreciation and inflation will have to be correspondingly higher. In short, there is always a combination of subsidies and nominal depreciation which ensures the constancy of real margins and whatever target for the real exchange rate the authorities decide to pursue.

We note that real depreciation is a desirable target in this context. The subsidy required to keep real margins constant is being financed at the expense of disposable wage income, which may very well induce a fall in demand and offset the virtuous supply-side effect of subsidies. Real depreciation serves to substitute consumption with exports and therefore helps to sustain the level of demand and investment (at unchanged real margins). Once more, we stress that, to achieve the desired outcome, what is required in principle is a *one-shot* change in the nominal exchange rate and in prices (and a one-shot real depreciation, if the latter is desired). This is important, not because the engineering of a one-shot price rise is a realistic possibility, but because it shows that the story does not require complete money illusion for gross wages. It only requires persistent money illusion for gross wages. It only requires persistent money illusion for the net wage. It is sufficient to perform the trick once, even if wage earners react later by requiring some form of indexation for the net wage, which they can do by asking the government to index the tax rate. The only thing that matters is that such indexation should be not applied retroactively (3).

2.2.2 The Strategy in Italy

We are well aware that reality is far more complicated than any simplified parable can tell. First, as expected, the true story features not a once-and-for-all jump in prices but persistently high inflation. Second, even though inflation helps the budget, it is hard to think that

---

(3) Another problem may arise from the presence of a Tanzi-Olivera (more correctly Bresciani-Turroni) effect, whereby tax collection occurs with a long lag behind earned income so that the real value of tax revenues decreases in the interim as prices rise. The latter, however, concerns situations of persistent hyperinflation, not a one-shot price increase.

policy-makers consciously plan inflation to raise extra revenue. Extra revenue is a welcome bonus, but monetary accommodation is likely to find other justifications. Third, by assuming that the tax system is the only source of monetary non-neutrality in the economy, we have neglected more traditional channels, like price stickiness. Finally, even within the period considered in this section, changes in unions' behaviour, as well as important differences in the strategies of enterprises, were certainly relevant.

Still, the purpose of parables is not to provide an exhaustive picture of reality, but to draw attention to one of its many relevant aspects, and we believe that our parable captures a crucial feature of the Italian recovery of the 1970s. Table 5 provides evidence on the

TABLE 5

ITALY: THE WEDGE AND THE COST OF LABOUR IN MANUFACTURING

	1977-1980	1981-1985
Real gross wage per unit of output (a) (% p.a.):		
(o.n.a.)	0.1	0.1
(n.n.a.)	1.3	−0.3
Real labour cost per unit of output (a) (% p.a.):		
(o.n.a.)	−2.2	0.7
(n.n.a.)	−1.2	0.1

	1976	1977	1980	1985
Wedge (b) (% of gross wage):				
(o.n.a.)	47.5	38.2	33.9	38.1
(n.n.a)	53.6	43.3	39.1	41.0
Social security contributions paid by employers (% of gross wage):				
(n.n.a.)	39.3	34.4	29.7	31.8

(a) o.n.a. data corrected for workers on subsidized layoffs and all data adjusted for self-employed; value added level and deflator at market prices; (b) ratio of the difference between labour costs and gross wages (mostly funds for severance indemnity and social security contributions paid by employers) to gross wages.

Sources: ISTITUTO CENTRALE DI STATISTICA: *Annuario di contabilità nazionale*; BANCA D'ITALIA: *Annual Report*; CENTRO EUROPA RICERCHE, *Rapporto*, n. 6, 1987.

reduction of the wedge between the gross wage and the cost of labour in the period under consideration. Before 1976, and after 1980, the cost of labour for enterprises rose slightly faster than the gross wage. Between 1976 and 1980, however, the (average) annual growth of labour cost was far below that of gross wages; so that real unit labour costs could fall. The reduction of the wedge was the result of a generous cut in social security contributions paid by enterprises and of an agreement with the unions to remove indexation from the funds set aside for severance indemnity (indexation of these funds was subsequently reintroduced in 1982. On the relevance of the two causes of reduction of the wedge in this period, see Banca d'Italia [2]). The ratio of contributions to gross wages fell by ten points between 1976 and 1980.

In spite of subsidies, the budget deficit as a proportion of GDP did not increase in the period: in fact, revenues rose more than expenditures net of interest and the primary deficit actually fell between 1974 and 1980. We are not implying that a budget problem did not already exist at the time. The size of the deficit was already a prominent issue, to which we shall return in section 4. We only maintain that the dynamics of revenues were not affected by the cut in social security contributions, as the rise in direct taxes provided plenty of room for the subsidies. Table 6 shows that the major source of rising revenues was the joint operation of steeply rising marginal tax rates — introduced with the 1974 tax reform — and of high inflation; fiscal drag was responsible for a rise of almost 9 points in the tax

TABLE 6

TAXES ON LABOUR INCOME:
CHANGES IN DIRECT TAX RATIOS ON EMPLOYERS' INCOME
(% of gross wage)

	1974-1977	1977-1980
Total	+2.8	+4.6
due to:		
rise in real income	+1.3	+0.4
discretionary changes	−2.5	−0.7
fiscal drag	+4.0	+4.9

Source: CENTRO EUROPA RICERCHE: *Rapporto*, n. 4, 1984.

burden on industrial workers between 1974 and 1980. Only a fraction of this was offset by discretionary measures of opposite sign. The inflation-induced increase in the average tax rate was such that the wedge between the cost of labour for enterprises and take-home pay actually increased in the period, in spite of the substantial cut in contributions.

There was in short a redistribution from wages to profits in industry by means of increased taxation of labour income which was induced by inflation rather than legislated by Parliament. Inflationary accommodation provided a way to offset the consequences of the nominal wage shock. The unions were not wholly blind to what was happening, but the compensating discretionary measures which they managed to obtain never affected the structure of marginal tax rates, which was the source of the fiscal drag. They only became fully aware of the issue in later years, when "restitution of fiscal drag" became a major bone of contention with the government. The structure of income taxation remained unchanged until the early 1980s when it was made less progressive, and deductions for employees' income were repeatedly increased. Today a repetition of past history, involving the use of inflation to raise tax revenues, would no longer be possible (in 1988 the government agreed to a *de facto* indexation of personal taxation by pledging to cut taxes on labour income so as to offset the previous year's fiscal drag in excess of 2% inflation). Learning by the unions took a long time, however, and that time was sufficient to undo the effects on the distribution of disposable income of the nominal wage shock (4).

The fall in real unit labour costs in 1977-1980 was greatly helped by the reduction of the wedge, and hence by subsidies. The acceleration of productivity growth (Table 4) was at least as important. There

---

(4) The consequences of inflation were not confined to those on tax revenues, as inflation also provided a powerful help to enterprises by eroding the real value of their outstanding debt contracted at fixed nominal interest rates. A measure of this effect is the size of the inflation correction of the reported net liabilities of the enterprise sector, which also shows the redistribution from households to firms and which in the period under consideration was as high as 5% per year on average (MINISTERO DEL TESORO [23]). Another measure is the difference between gross operating margins and margins net of *real* interest payments: the latter's rise is far steeper than that of the former (BODO - VISCO [7]).

are certainly several factors accounting for the behaviour of productivity. In 1977 the unions agreed to remove some restrictive practices and to allow a more efficient use of equipment. Still the resulting increase in flexibility did not remove the most relevant obstacles to firings and lay-offs. This is why there is little doubt that a recovery of demand and output was a necessary condition for faster productivity growth and for exploiting the new machinery installed during the 1973-1974 investment boom. The recovery was led by exports which grew at an average real rate of over 9% between 1976 and 1980.

Graph 3 shows the real effective exchange rates of the lira computed using manufacturing wholesale prices and unit labour

GRAPH 3

ITALY: REAL EXCHANGE RATES
(indices: 1975 = 100)

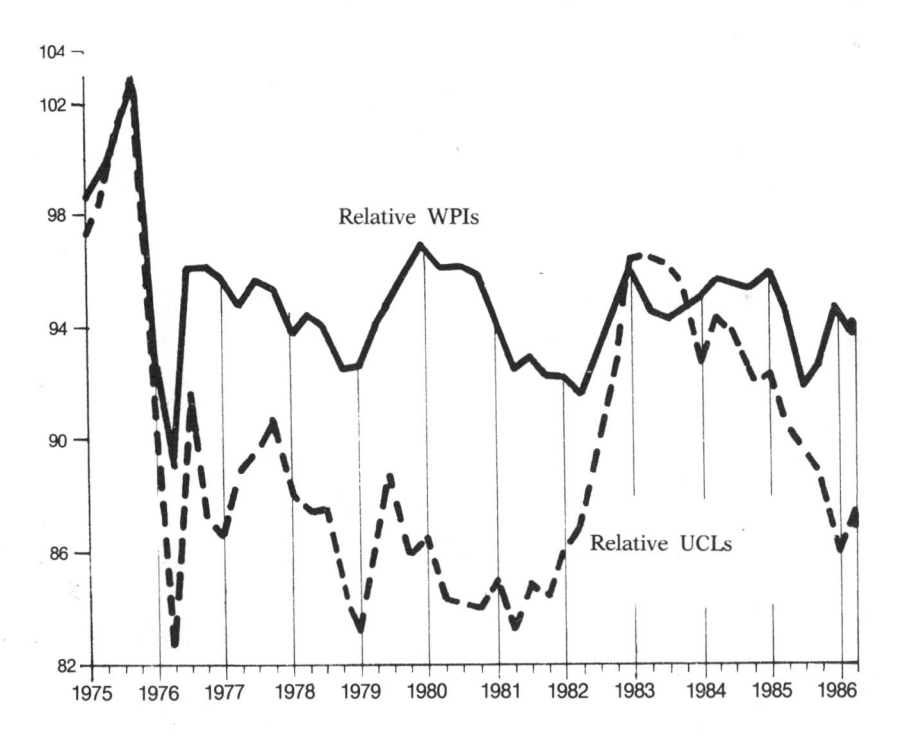

Source: IMF: *International Financial Statistics, Supplement on Price Statistics*, 1986.

costs. Between 1975 and 1978, the nominal depreciation resulted in a large real depreciation. The trend reversed in 1979, when Italy joined the EMS. The gap between the real exchange rate based on wholesale prices and that based on unit labour costs widened until 1981, providing further evidence of the recovery in margins. Table 7 compares the relative export performance (growth of exports minus growth of export markets) in Italy and in the UK. The difference between the two countries, as well as the strength of Italy's export drive in 1977-1979, is remarkable. The boom in domestic demand and real appreciation account for the turn-around in export performance in 1980. A large part of the explanation for the difference between the British and Italian experiences can be traced back to the evolution of their effective real exchange rates. Graph 4 illustrates the large appreciation of sterling which is in sharp contrast with the quasi-stability of the lira during the period 1976-1981.

To what extent was the fall of the lira a conscious decision by the Italian monetary authorities? The answer is different for the two episodes of sharp depreciation, namely the first half of 1976 and 1978. In 1976 the fall of the lira was not planned; it resulted from one semester of extravagantly expansionary monetary policy, which caused accelerating inflation and a deterioration of the current balance. Speculation against the lira then forced the authorities to close the exchange markets and to turn to the IMF for a stand-by arrangement (which was never used). In contrast, by the end of 1977 the current account was in balance and there were massive capital

TABLE 7

RELATIVE EXPORT PERFORMANCE
OF ITALY AND THE UK (*)

	Italy (o.n.a.)	UK
1976	+0.9	−2.0
1977	+2.6	+3.0
1978	+5.0	−4.5
1979	+0.4	−5.0
1980	−4.7	−8.5

(*) growth rate of exports minus growth rate of export markets for manufactures.
Sources: OECD, *Economic Outlook*, various issues; and ISTITUTO CENTRALE DI STATISTICA.

THE REAL EXCHANGE RATE IN A DISINFLATION
(relative unit labour costs)

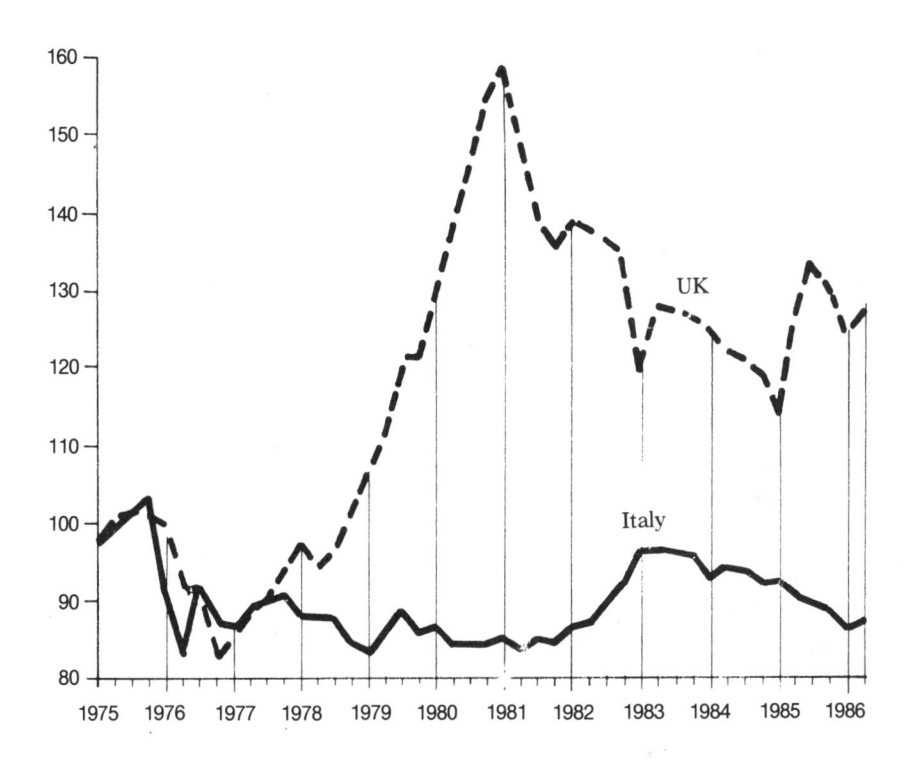

GRAPH 4

Source: IMF: «International Financial Statistics», *Supplement on Price Statistics*, 1986.

inflows. Not only was the exchange rate not allowed to appreciate but, on the contrary, the authorities pursued a deliberate policy of real depreciation inspired by several motives. First, there was the need to replenish foreign exchange reserves. Second, the fall of the dollar at the end of 1977 provided an opportunity to engineer an appreciation *vis-à-vis* the major import currency, and a further depreciation against the stronger currencies of Italy's export markets in Europe. Third, the Italian monetary authorities, having more fears than enthusiasm for the EMS, thought it wise to set some more depreciation in store during the second half of 1978 and in early 1979.

The wisdom of this policy of deliberate depreciation has later been explicitly or implicitly questioned on the grounds that stricter exchange rate discipline would have lowered inflation without affecting the growth performance in the long run (Andreatta and D'Adda [1]). The Governor of the Bank of Italy faced this issue at that time and gave the following answer: «Though aware of its role in shaping the dynamics of prices, we guided the external value of the lira so as to permit a growth of exports laying the foundations for a recovery of accumulation and of employment less conditioned by the external constraint. (Banca d'Italia: *Annual Report*, 1979)».

We agree with the Governor. Depreciation and subsidies were part and parcel of the same strategy, designed to pull the economy out of the low growth-rising costs trap into which it had fallen with the oil shock. Real wage flexibility was perhaps a first-best solution, but his perfect course of action was followed nowhere in Europe. In view of her growth and investment performance, the Italian second-best solution does not appear to be inferior to the strategies followed in other countries. Cost and price inflation was lower in Germany, but so was productivity and GDP growth. Cost inflation was slightly better in France and the UK, but profit margins were squeezed in both countries, because of lower price inflation in France and because of sluggish output and productivity in the UK. In Italy, the early recovery in profit margins and in demand sparked off an investment boom in industry, which lasted for two years (5). In contrast, in the UK, a country similar to Italy in conditions prevailing in the labour market,

---

(5) The Governor of the Bank of Italy provided a vivid picture of the boom in his *1981 Report:* «First exports, then private consumption pulled the economy out of the recession in the course of 1978. Consumption kept growing in the first half of 1979. [...] The dynamics of exports and consumption sparked a strong recovery of investment, especially in machinery and equipment, in the second half of 1979. [...] For nine months since the beginning of the expasionary phase gross domestic product grew at an annual rate of 10%.

The nature of investment varied with the size of firms (BARCA - MAGNANI [5] [6]; BARCA [4]; HEIMLER - MILANA [22]. Scrapping and disinvestment were particularly important for larger enterprises. The analysis of a Bank of Italy sample (BARCA - MAGNANI [6]) shows that between 1977 and 1982 there occurred a remarkable shortening (2.5 years) in the average life of equipment. This matches some evidence of a younger average age of equipment in Italy than in other countries (BANK OF ENGLAND [3] and CHAN-LEE - SUTCH [12]).

the strategy was to let the hardships of recession and unemployment take care of the problem: growth and investment did eventually resume, but, as we shall stress later, at greater cost in terms of output and capacity.

3. - The Output Cost of Disinflation

With 1980: «there came to an end a two-year period in which the real growth rate of GDP and private consumption neared 10%, that of investment in machinery and equipment 30% ...: such ratios are twice or three times as high as in the OECD area. The Italian economy surpassed all other countries in the pace of growth of income and even of employment, but paid the price in terms of external imbalances and inflation» (Banca d'Italia: *Annual Report*, 1981).

To put fiscal non-neutrality to work and pay for the subsidies, a jump in the price level is in principle sufficient; but inflation is seldom a stable process: the rate of price increase accelerated from 11% in the Winter of 1979, to 25% in the Fall of 1980. More importantly, considering that in March 1979 Italy had joined the EMS, the inflation differential relative to Germany widened from 12% in 1979 to 16% in 1980. One might thus be tempted to conclude that the output cost of reducing real wages had only been postponed and that waiting made things even worse; indeed by 1980 not only was high inflation built into expectations and the wage-price mechanism, but the real wage problem was coming right back as unions realized to what extent they were cheated by fiscal drag. What comes as a surprise, however, is the mildness of the Italian recession during the disinflation. Between 1980 and 1987 the inflation differential relative to Germany fell from 16% to 4.5%, but the economy kept creating jobs. It is instructive to compare the path of prices and output in the British and Italian disinflations. The amount of disinflation was almost the same: there is, however, a difference in terms of the timing and the output cost. A similar reduction in the inflation differential *vis-à-vis* Germany took three years in the UK but twice as long in Italy. The output cost, however, was very different.

TABLE 8

THE OUTPUT COST OF DISINFLATION (*)

	UK 1980-1983	Italy 1980-1987
Reduction in the (CPI) inflation differential relative to Germany:		
(a) percentage points	10.0	9.0
(b) % of the initial level	84.4	67.2
(c) cumulated increase in unemployment	21.2	17.6
(d) cumulated output loss	11.7	5.6
Sacrifice ratios:		
unemployment measure (c)/(a)	1.96	1.96
(c)/(d)	2.45	2.62
output gap measure (d)/(a)	1.08	0.62
(d)/(b)	0.14	0.08

(*) Cumulative output loss: sum of deviations from trend GDP growth (log-linear trend: 1970-1987); cumulative increase in unemployment: sum of year to year changes in OECD standardized unemployment.

Sources: OECD: *Economic Outlook*, except unemployment for Italy from BANCA D'ITALIA, *Annual Report*, 1988.

Model-free estimates of the output cost of disinflation, like "sacrifice ratios", are often difficult to interpret because they lack a theory predicting what the path of output and employment would have been in the absence of monetary contraction. We present in Table 8 two different measures of this cost in Italy and in the UK. The first set of numbers measures how much unemployment it takes to bring inflation down by 1% (6). An alternative measure uses output gaps, computed as deviations of real GDP from its (linear) trend growth rate between 1970 and 1987. For each of the two countries the interval over which the sacrifice ratio is computed ends when the inflation differential relative to Germany stabilizes. We show two different

---

(6) For example a 10% reduction in inflation that is achieved in four years, and is associated with an increase in unemployment of 0.5% per year (say from 5% in year 1 to 5.5% in year 2, to 7% in the last year), corresponds to a sacrifice ratio of $1/2 = (0.5 \times 4 + 0.5 \times 3 + 0.5 \times 2 + 0.5 \times 1)/10$. This measure of the sacrifice ratio is computed for example in SACHS - WYPLOSZ [28].

measures of the reduction in the inflation differential (and therefore also of the sacrifice ratios), namely the absolute fall in the differential, and its percent reduction relative to the initial level. The latter accounts for the fact that squeezing inflation by 1% is harder at single-digit inflation levels than it is at double-digit inflation. The output cost was lower in Italy. The superior performance of Italy emerges from the output gap measure. This is in our opinion a better measure, since cumulated unemployment (for which Italy did not do better than the UK) tends to penalize countries, such as Italy, where disinflation is slower, by attributing the persistence on unemployment to disinflation. The difference would be much larger if the reference were actual output growth rather than the deviation from trend, as trend growth is 3.1% in Italy, and 1.7% in the UK.

But why was disinflation relatively costless? We first deal with two possible answers. First, the EMS, by shifting expectations, has improved the output-inflation tradeoff. Second, the government has actively tried to reduce the sacrifice ratio, a view based on the observation that between 1980 and 1985, out of 535 thousand (net) new jobs created in the entire economy, 400 thousand were new government jobs. In our opinion both interpretations overlook the supply side of the story, which is crucial to account for the low output-cost of Italian disinflation.

3.1 Credibility and the EMS

Bringing inflation down requires a change in inflationary expectations on the part of price setters. To convince price setters that an announced contraction will be lasting and credible, and gain reputation, monetary authorities can proceed in two ways. The first is to show that, even in the depths of a recession, the announced monetary targets are not reneged on. The recession will come because the monetary contraction necessary to disinflate is imposed on an economy where inflation expectations are high, and because the very fact that the monetary authority sticks to the announced contractionary path comes to private agents as a surprise. Alternatively, monetary authorities can seek to influence expectations with some institutional refotm, such as a change in the exchange-rate regime (7). Can the transition from flexible to fixed exchange rates bring about an improvement in the output-inflation tradeoff, and facilitate the disinflation effort? Suppose a country decides to peg its exchange rate passively to another country, whose monetary authorities enjoy a reputation as credible inflation-fighters. By a 'passive peg' we mean that the monetary authorities, after announcing the exchange-rate parity, simply accomodate the other country's monetary policy, without attempting to influence its choice of targets. In the private sector, wage and price setters will appraise the credibility of this institutional reform in terms of the probability they assign to the consistent pursuit by the authorities of the announced exchange-rate target. If, and only if, the target is a credible one, expectations will adjust and the process of disinflation will be eased.

The argument that joining the EMS has helped Italy in its disinflation efforts of the 1980s rests crucially on the assumption that exchange-rate targets are more credible than monetary targets. The way to assess the empirical relevance of this argument is to show that the decision to join the EMS has produced a shift in expectations. The empirical evidence reported in Giavazzi and Giovanini [16] is consistent with this view but suggests a long lag between the start of the EMS and the effect on expectations: in Italy the shift in expectations occurred in the first quarter of 1985, six years after the start of the EMS; in France in March 1983; in Ireland in the Fall of 1982. The timing of these shifts suggests that the turn-around in expectations was induced by some specific set of measures which signalled that there had been a change in policy regime: the turn-around in macroeconomic policies of the first Mitterand government in France, and a similar policy turn-around in Ireland in the summer of 1982 (the French and Irish experiences are discussed in Sachs and Wyplosz [28], and Dornbusch [14]), respectively). As for Italy, the government set, by decree, a ceiling to wage indexation limited to one year, 1984. By itself this measure would have a very small effect on inflation. But

---

(7) The arguments discussed in this section are developed more fully in GIAVAZZI - GIOVANNINI [16] and [17] chap. 5) where the empirical results to which we refer are also reported.

the opposition and the more militant unions called for a national referendum and were defeated. This defeat, and an unusual display of firmness on the part of the government, affected expectations far more than the measure itself (for an account of these events and for an estimate of the shift in expectations see Gressani et Al. [21]). EMS membership might have helped by providing a justification for unpopular policies, but the data strongly suggest that the new exchange-rate regime did not automatically produce an improvement of the output-inflation tradeoff. Governments had to prove that they were prepared to bear the cost of unpopularity before price-setters became convinced that the commitment to the new monetary targets was lasting.

3.2 *Fiscal Accomodation*

Disinflation has important implications for the budget. It reduces the portion of the deficit that can be financed by printing money and it raises real interest rates which adds to the cost of servicing the public debt. Disinflation therefore requires a shift not only in monetary policy, but also in fiscal policy: failure to adjust fiscal policy to the new monetary conditions will ultimately result in a debt problem. Accompanying disinflation with fiscal contraction, on the other hand, may worsen the recession already induced by the slowdown in money growth which, in the short run, may well worsen the budget deficit despite the improvement in the structural budget balance.

TABLE 9

CHANGE IN STRUCTURAL BUDGET BALANCE
(% of GDP) (*)

	1980	1981	1982	1983	1984	1985	1986
UK	+1.1	+2.9	+1.4	−1.3	−0.5	+0.5	−0.3
Italy (o.n.a.)	+1.0	−3.4	0.0	+2.2	−0.3	−0.9	+0.6

(*) A positive sign indicates a move toward fiscal restriction.
Source: OECD: *Economic Outlook*.

Between 1980 and 1982, Britain accompanied her disinflation by a sharp budgetary contraction. In Italy, on the contrary, there was no adjustment in fiscal policy. The difference is illustrated in Table 9. Between 1980 and 1982 the UK structural balance improved by 5.4% of GDP, while in Italy it worsened by 2.4% (by 1986 the primary deficit ratio was still at its 1980 level). The path of fiscal policy may have affected the output cost of disinflation; in Italy, disinflation was preceded by a sharp fiscal expansion, between 1980 and 1981, only partly reversed in the following period. (Between 1979 and 1981 indexation for public sector salaries was increased, pensions were raised and so were tax deductions and family allowances, while generous increases were granted to civil servants with the payment of substantial lump-sum advances.) This expansion, superimposed on a situation already characterized by excess demand, raised the inflation peak and caused a large external imbalance. This initial fiscal impulse certainly contributed to sustaining demand, but we have no estimate of the size of its contribution. It is likely that the output cost of disinflation would have been higher if Italy had followed the path of British fiscal policy. But the crucial question is the extent to which this policy contributed to the build-up of public debt, which in the long run is the true cost of delaying adjustment by means of fiscal expansion. We address this question in section 4.

3.3 *The Timing of the Supply Squeeze*

In order to break inflation inertia it was necessary to convince firms that they could no longer rely on a depreciating exchange rate and on subsidies: to preserve their profit margins thay should now cut costs and exploit to the best their earlier investments. At the macro level the change was signalled by a tightening of monetary policy and a sharp rise of interest rates. The authorities let the lira appreciate substantially: realignments in the EMS, though frequent, always fell short of the cumulated inflation differential (8). The generous sub-

---

(8) The rise of the dollar mitigated the rise of the real effective exchange rate which was half as large (+7%). For an analysis of the role of a real appreciation in enhancing the credibility of a disinflation, see GIAVAZZI - PAGANO [18].

sidies which had lowered the cost of labour earlier now came to an end: the wedge between the cost of labour and the gross wage increased again (Table 5). Help to industry did not, however, come to an end. Subsidies to reduce the cost of labour were replaced by provisions for firms planning to increase productivity by rationalizing, and reducing, the use of labour. State financing of layoffs (*Cassa integrazione*) due to restructuring and early retirement schemes encouraged labour shedding, and made it possible to bypass the staunch opposition of the unions to outright firing of industrial workers. Between 1980 and 1984 the number of hours paid through this system (at 80-90% of the ordinary wage) increased by three times. Employment in larger enterprises, which made far greater use of the scheme, fell by more than 21%, and yet the number of hours lost through strikes declined by 80%, from 75 million hours lost in 1980 to 16 million in 1985.

The attitude of firms changed. The unions changed too, threatened by sticks (the decree that partly suspended wage indexation and the layoffs which discriminated against the more militant workers) and lured by a few carrots (a clumsy but not ineffective version of tax-based incomes policy, whereby tax concessions to offset fiscal drag were granted in exchange for wage moderation). The result was far greater flexibility in the use of the work force and fast increase in prductivity. This policy was effective, but its effectiveness depended crucially on the fact that it came *after* the earlier recovery in margins, investment boom and modernization of equipment.

Once exchange-rate accommodation came to an end and competitiveness started to fall, profit margins were squeezed. But as the squeeze came at a peak of the profit rate, there was room to reduce margins without turning them negative. The outcome strong pressure on firms to adapt to the new monetary regime, but few bankruptcies and plant closures. This, in our opinion, is the explanation for the low output cost of the Italian disinflation. In the UK, at the start of the disinflation, profit margins were at an all-time low. As the pressure from the real appreciation mounted, along came bankruptcies and plant closures: these are partly irreversible decisions that imply permanent dissipation of physical and human capital.

The role of the initial level of profits in determining the output cost of the disinflation provides a good example of hysteresis — namely of the possibility that temporary fluctuations may have long-lasting effects on the economy. This point, which to the best of our knowledge has never been documented, obviously deserves more careful empirical investigation. Yet, the lesson seems to be that exerting pressure on the supply side may be an effective way to speed up the adjustment of prices and wages — in the sense that it may reduce its output cost — provided it comes at a time when firms are ready to bear a squeeze on profits.

4. - Is Debt the Price?

In the mid-1970s the ratio of public debt to GDP was below 50%; it has now passed the 100% mark. High debt levels are not unusual in Europe. But contrary to other countries where debt is also high — Belgium for example — Italy still runs primary deficits close to 3% of GDP: there is as yet no sign of a primary budget surplus sufficient to service and stabilize the debt. Even though Italian government debt is almost entirely held domestically, so that a problem of solvency in the strict sense does not exist, the fear of financial instability is widespread, as signalled for example by the inability of the Treasury to lengthen the maturity of debt. (On the management of the Italian debt see Giavazzi and Spaventa [19]) The obvious question is whether the smart policies of the 1970s and early 1980s have simply postponed the bill, that now falls due in terms of an unsustainable path for public debt.

Italy has known high debt levels before. Since 1861, the date of birth of the Italian state, the ratio of public debt to GDP has grown beyond 100% three times already: in the early years of the new nation (1861-1910), and at the time of the two world wars (see Graph 5). The three episodes of debt stabilization that followed each surge of the debt ratio, correspond to three different ways of reducing a high debt. Between 1900 and 1910, it was the rapid growth of real income which lowered the debt to GDP ratio. After each war, on the contrary, debt was reduced through some form of repudiation: a forced consolidation in 1926, and inflation in 1946-1947. After World War II there

Italy: the Real Effects of Inflation and Disinflation 47

GRAPH 5

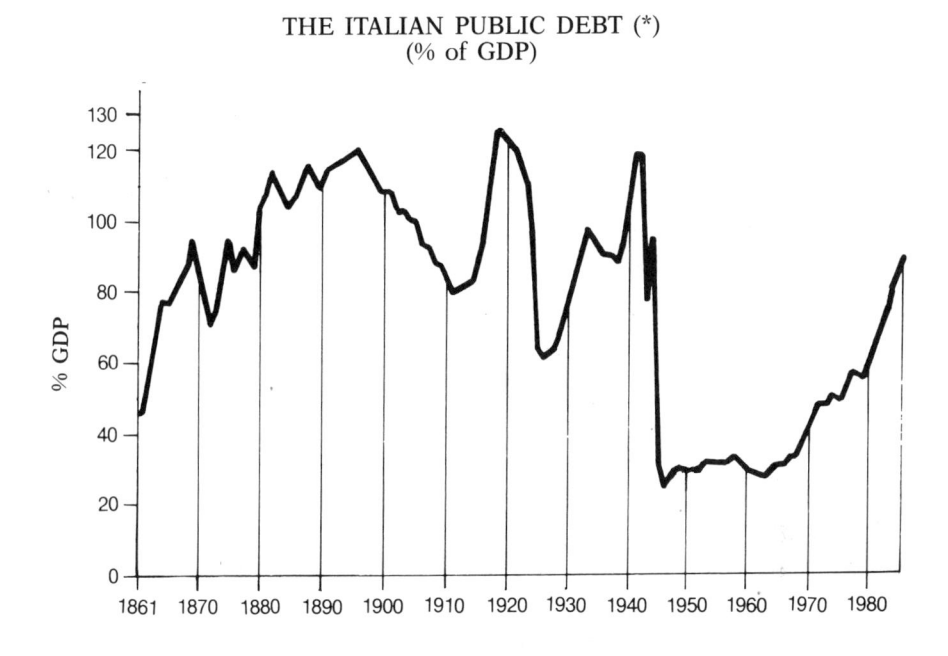

THE ITALIAN PUBLIC DEBT (*)
(% of GDP)

(*) Total public sector debt including debt held by the Central Bank.
Source: MINISTERO DEL TESORO, 1988.

was a 20-years period of debt stability which ended in 1969. In only three years, between 1970 and 1973, the ratio of debt to GDP jumped from 33% to 50%. These are the years when Italian public finances went out of balance. Never since have they recovered. The most recent growth of the debt to GDP ratio occurred in a period of peace and is thus more similar to the 1880-1910 episode. It is interesting to note that the only case of debt growth which has not resulted in some form of repudation is precisely this peacetime 1880-1910 episode.

In order to understand what happened to the budget in the early 1970s we decompose in Graph 6 the growth of public debt into its two components: the primary deficit and debt service. In the 1960s, stability of the debt level was guaranteed by a combination of moderate primary deficits and interest rates lower than the growth rate of the economy. Between 1970 and 1973, the primary deficit

GRAPH 6

PRIMARY PSBR AND DEBT SERVICE (*)
(% of GDP)

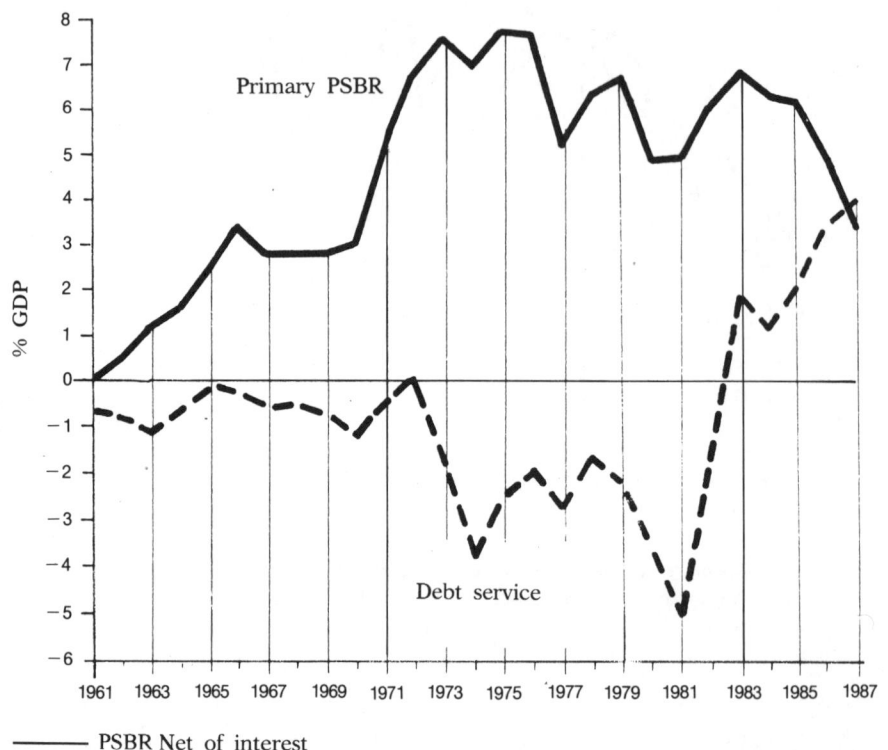

——— PSBR Net of interest
- - - - (i-n)b

(*) The measure of the primary deficit reported in Graph 6 is the Public Sector Borrowing Requirement net of interest. The contribution of debt service to the growth of public debt is $(i-n)b$, where i is average nominal interest rate on public debt, n is the growth rate of nominal GDP, and b is the ratio of debt to GDP.

Source: MINISTERO DEL TESORO, 1988, and authors' calculation.

increased from 4.3% to 8.3% of GDP: this increase was not cyclical (1973 was a boom year with 7% real growth), but was caused by a structural jump in public expenditure not matched by a corresponding change in revenues. The early 1970s were a period of big social reforms: extension of the years of compulsory schooling, reform of the health-care system, the decision to link pension benefits to earnings, rather than to contributions, etc. The gap that those social bills opened in public finances, documented in Table 10, has never since been closed. In the mid-1970s the tax base was widened through a major tax reform. Between 1973 and 1985 the combination of a larger tax base, higher tax rates and inflation raised revenues from 29% to 41% of GDP: enough to cover the increase in expenditure over the same period (also approximately equal to 12% to GDP), but not enough to close the gap opened in the earlier period. Thus, for 15 years (1973-1985) the primary deficit has fluctuated between 5% and 8% of GDP.

TABLE 10

THE SOURCE OF THE ITALIAN DEBT PROBLEM
(public sector, % of GDP)

	1970 level	1970-1973 change
Revenues	29.2	−0.5
Expenditure (net of interest)	33.5	+3.5
purchases of goods and services		+1.2
wages and salaries		+0.4
pensions and other social transfers		+1.3
other items	3.3	+0.6

Sources: MINISTERO DEL TESORO, 1988.

Once a gap between expenditures and revenues is opened, the subsequent growth of the debt ratio depends on the real interest rate and on the growth rate of the economy. As shown in Graph 6, just when primary deficits increased, inflation turned the stock of debt into an asset rather than a liability, as nominal interest rates fell much below the growth rate of income and helped to stabilize the debt ratio. Though low real rates were not special to Italy in the 1970s, there is evidence that exchange controls allowed the Italian authorities to keep domestic rates below the level they would have reached otherwise. Between 1974 and 1983 onshore rates were on average 350 basis points lower than the corresponding offshore rates. The ability to

GRAPH 7

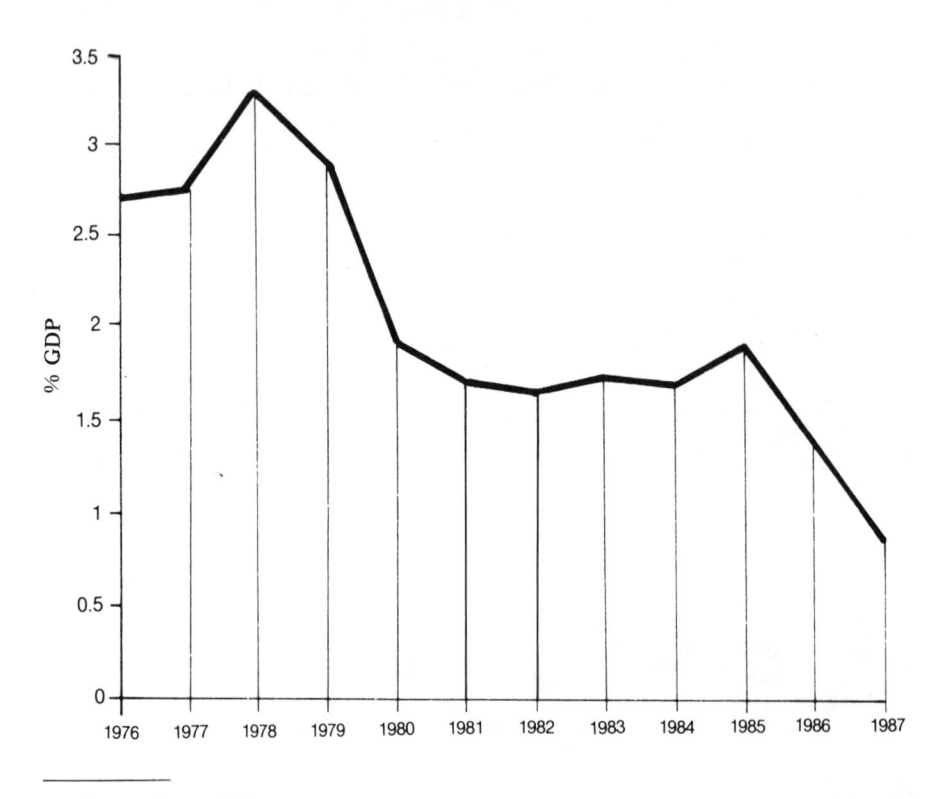

SEIGNIORAGE
(% of GDP)

Source: PAGANO [27].

raise revenue through the seigniorage attached to money creation is another way to slow down the growth of marketable debt. As shown in Graph 7, seigniorage revenue contributed 2-3 percentage points of GDP per year throughout the late 1970s (9).

Graph 8 provides a rough estimate of the path of the debt ratio under an alternative story of greater monetary virtue in the 1970s and less fiscal profligacy in the early 1980s. Had Italy followed a low-infl-

---

(9) Most of this revenue, as discussed in GIAVAZZI [15] is accounted for by the level of bank reserves, much higher in Italy than elsewhere in the industrial countries.

Italy: the Real Effects of Inflation and Disinflation   51

GRAPH 8

ALTERNATIVE DEBT PATHS (*)
(difference from actual debt as % of GDP)

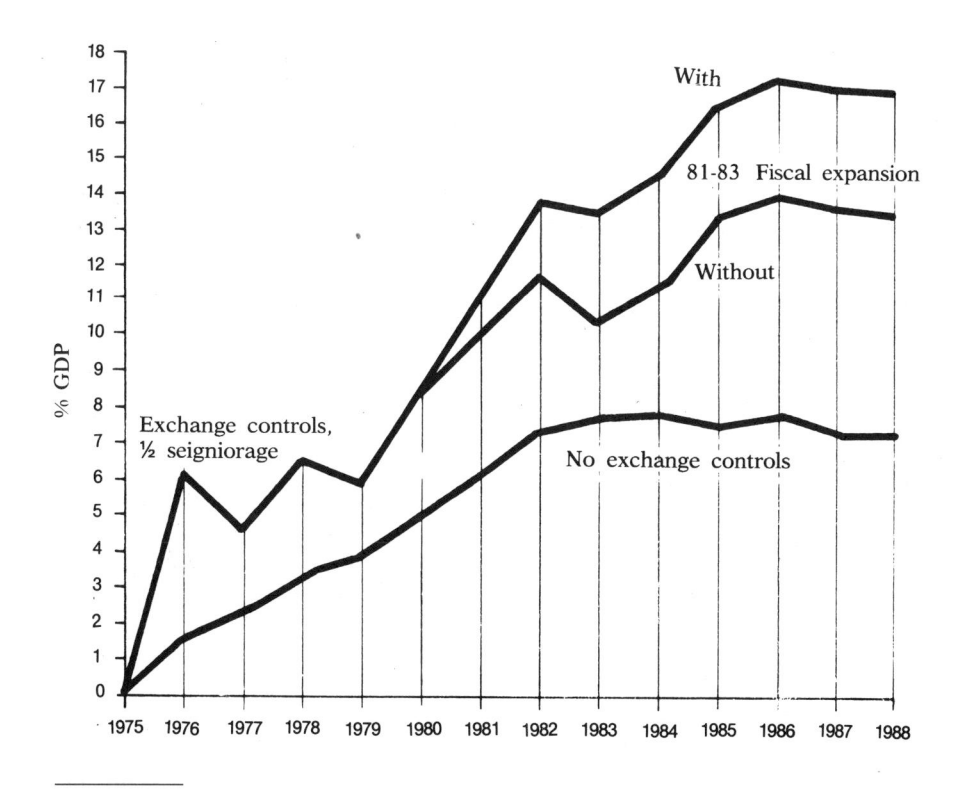

[Graph showing three curves from 1975 to 1988, labeled "With", "81-83 Fiscal expansion", "Without", "Exchange controls, ½ seigniorage", and "No exchange controls"]

(*) In these simulation 'debt' is the marketable debt of the state sector.

ation path since 1976, domestic interest rates would have been at the (covered) level of international rates: the lower line is drawn by simply adding the corresponding extra debt service cost measured by the difference between the offshore and the onshore interest rates. If we also assume that seigniorage was kept at only 1% of GDP per year over the period, we obtain the upper line. To take care of the fiscal expansion in 1981, we have subtracted 1% of GDP from primary deficits in the years 1981-1983: adding this assumption to the other two, we obtain the middle line. In the three simulations, we use the actual nominal income growth rates and primary deficits (except in

1981-1983 for the third case). These simulations show that the policies discussed in this paper were not by themselves responsible for the growth of debt and that more conventional policies would actually have been more costly in terms of debt.

In the last few years, the safety nets which sterilized the effects of fiscal imbalances on debt in the 1970s have been removed: the need to set monetary policy consistent with German targets as well as financial liberalization have both cut into seigniorage and made it impossible to control real rates. Italy now faces the effects of the fiscal imbalance created in the early 1970s Revenues will have to increase: this, however, is the delayed price to be paid for the social reforms of the early 1970s. It is not a bill to settle either for the supply-friendly policies of the late 1970s, or for the policies which later reduced the cost of disinflation.

5. - Conclusions

In the troubled period of the 1970s, characterized by several shocks and a complete lack of wage flexibility, recourse to conventional policies would have implied a prolonged period of depression and postponed supply-side adjustment. Less inflation and greater exchange-rate stability would have been the prize, but a costly one in terms of output, and particularly in terms of investment. Instead, the Italian response to the oil shocks was at odds with this conventional wisdom. Inflation was used to give government the means to boost profit margins. Despite a fairly complete system of wage indexation, inflation worked through the non-neutrality of the income tax system, and some degree of real-wage myopia of the trade unions. That route could, however, not be pursued for too long because its costs were rising and its benefits were declining, particularly as inflation almost went out of control. EMS membership marked the watershed and precipitated the change, but a change was unavoidable anyhow. Disinflation was not achieved by conventional means either, as the change in monetary regime was not accompanied by a consistent shift in fiscal policy. It might have been expected that the retribution for profligacy would be high. This was not the case, however. The earlier recovery of margins and the investment boom eased the consequences of monetary and exchange rate discipline; with high profit margins and the new capital stock installed, pressure could be exerted on industry to cut costs and to adapt quickly to the new regime without undue sacrifices in output. As of 1989, Italy has solved most of its macroeconomic problems, save one. It is our view, however, that the debt problem does not originate in the particular sets of policies reviewed in this paper. The primary budget deficit was promptly and widely increased between 1970 and 1973, and never reduced since. Yet, it has not further deteriorated either; the explosive growth of the debt to GDP ratio is a consequence of this early increase in the primary deficit, combined with high real interest rates. Undoubtely, the primary deficit must now be turned into a sizeable surplus and this may well provoke some severe macroeconomic hardship: this, however, will be the price to pay for the 1970-1973 period, and not for the policies discussed in this paper.

The Italian experience raises a number of questions about the timing of stabilization policies after a major supply shock. One important result is that, in spite of indexation, inflation may be an effective policy instrument and that disinflation may be relatively painless. Timing seems, however, essential to success. The comparison with the British case is illuminating in this respect. With compressed profit margins, adjustment in Britain took partly the form of plant closures, thereby dissipating physical, and possibly human, capital. By boosting profit margins first, and subsequently imposing adjustment, Italy never underwent the massive wave of plant closures observed in the UK. Adjustment is a slow process as firms' strategies are limited by the speed at which labour markets can absorb large-scale restructuring: the timing and the sequence of policies are thus essential to make sure that temporary fluctuations do not have long-lasting effects on the economy.

APPENDIX

In this appendix we formalize the parable of section 3.1 in the text. Let m be nominal margins. ω is the nominal wage, v is the real price of materials. Nominal margins are equal to the price minus variable costs, so that:

$$p = \alpha w + m + \beta v p$$

We now define real wages and real margins respectively equal to:

$$\tilde{w} \equiv w/p \qquad \tilde{m} \equiv m/p$$

The economy is hit by a terms of trade shock equal to v. Wages are indexed. We shall consider two cases. In the first the economy trades only with the oil exporter at given relative prices, and wages are indexed to the price of domestic output. In the second the economy also trades with another industrial country, and wages are indexed to a basket of domestic and foreign final goods.

Case 1

As $\dot{w}/w = \dot{p}/p$, after an increase in the relative price of materials real margins fall by:

$$\dot{\tilde{m}} = -\beta \dot{v}$$

There is a progressive tax on labour incomes with elasticity $\eta > 1$. Nominal revenues per unit of output are:

$$t = \alpha \tau w_t^\eta, \qquad \eta > 1$$

Initially the government finances a given level of expenditure raising taxes on labour income by the amount:

$$\alpha \tau w_0^\eta$$

The next revenue per unit of output after a price rise is:

$$\alpha \tau \{[w_0(1 + \pi)]^\eta - w_0^\eta(1 + \pi)\}$$

where:

$$\pi \equiv \dot{p}/p.$$

The government now grants a subsidy s_t per unit of output in the form of a proportional contribution, σ, on the wage per unit of output:

$$s_t = \sigma \alpha w = \sigma \alpha w_0 (1 + \pi)$$

Let $\omega_0 + 1$, and let the net revenue from the price rise be fully devoted to subsidies: the contribution on the wage that can be financed is:

$$\sigma = \tau[(1 + \pi)^{\eta - 1} - 1]$$

Consider now a country with fixed real wages that wants to grant a subsidy such as to bring real margins back to their pre-shock level. We can compute by how much the price must rise to finance such a subsidy. The subsidy required is:

$$s = p\beta\dot{v}$$

i.e.:

$$\sigma = (\beta\dot{v})/(\alpha\tilde{w})$$

The price rise required to finance the subsidy is:

$$(1 + \pi)^{\eta - 1} = (1/\tau)[(\beta\dot{v})/(\alpha\tilde{w})] + 1$$

where:

$$\pi = (1/\tau)\,[(\beta\dot{v})/(\alpha\tilde{w})] \qquad \text{for } \eta = 2$$

Similarly we could compute price rises to finance any other recovery of margins, for example to bring real margins back to the level they would have been in the case of fixed nominal wages and nominal margins.

Case 2

Wages are now indexed to a basket of home and foreign final goods:

$$\dot{w}/w = \lambda\,(\dot{p}^*/p^* + \dot{e}/e) + (1 - \lambda)\,\dot{p}/p$$

where: p^* is the foreign currency price of foreign final goods, and e is the nominal exchange rate with the other industrial country. We shall assume for simplicity that $\dot{p}^* = 0$. The rate of inflation now is:

$$\pi = \frac{(\dot{m}/m)\,\tilde{m} + \alpha\lambda\tilde{w}\,(\dot{e}/e) + \beta\dot{v}}{\tilde{m} + \alpha\lambda\tilde{w}}$$

The change in real margins is:

$$\dot{\tilde{m}} = \tilde{m}\,(\dot{m}/m - \pi) = \frac{\alpha\lambda\tilde{w}(\dot{m}/m - \dot{e}/e) - \beta\dot{v}}{\tilde{m} + \alpha\lambda\tilde{w}}\tilde{m}$$

The change in real margins now depends on the change in nominal margins, so that there exists a rate of inflation such as to prevent a fall in real margins. That rate of inflazion, however, causes a real appreciation in the home country, for the dynamics of the real exchange rate are given by:

$$(\dot{e}/e) - \pi = \frac{\tilde{m}(\dot{e}/e - \dot{m}/m) - \beta\dot{v}}{\tilde{m} + \alpha\lambda\tilde{w}}$$

If $(\dot{m}/m) = (\dot{e}/e) + (1/\alpha\lambda\tilde{w})\beta\dot{v}$, as required to keep real margins constant, the real exchange rate will appreciate by $\beta\dot{v}/\alpha\lambda\tilde{w}$.

Subsidies are now used to restore profit margins preventing a real appreciation (or even engineering a real depreciation); the nominal exchange rate will have to move to provide inflation sufficient to finance the subsidies.

The real exchange rate will remain constant (or depreciate) depending on:

$$\dot{m}/m \leq \dot{e}/e - (\beta\dot{v})/\tilde{m}$$

Substituting this expression in the equation showing the dynamics of real margins, we obtain the subsidy required to keep real margins unchanged:

$$s \geq p\beta\dot{v}$$

The inflation rate is:

$$\pi \leq \dot{e}/e$$

where the inequality signs hold if the target is a real depreciation.

To finance the subsidies it is necessary that:

$$\sigma = \tau[(1/\pi)^{n-1} - 1] \geq (\beta\dot{v})/\alpha\tilde{w}$$

Consider for simplicity the case $\eta = 2$: then the rate of depreciation required to finance the subsidies is:

$$\dot{e}/e \geq (1/\tau)(\beta\dot{v}/\alpha\tilde{w})$$

By granting subsidies and suitably changing the nominal exchange rate it is always possible to keep real margins constant and achieve any desired value of the real exchange rate.

BIBLIOGRAPHY

[1] ANDREATTA N. - D'ADDA C.: «Effetti reali o nominali della svalutazione? Una riflessione sull'esperienza italiana dopo il primo shock petrolifero», *Politica economica*, n. 1, 1985.

[2] BANCA D'ITALIA: «Costi e profitti nell'industria in senso stretto: una analisi su serie trimestrali 1970-80», *Bollettino*, n. 36, January-December, 1981.

[3] BANK OF ENGLAND: «Trends in Real Rates of Return», *Quarterly Bulletin*, n. 3, 1988.

[4] BARCA F.: *Sviluppo e ristrutturazione delle imprese industriali: le due fasi: 1978-80 e 1981-85*, Banca d'Italia, mimeo, 1987.

[5] BARCA F. - MAGNANI M.: «Ristrutturazione e disinvestimento anticipato nella medio-grande industria italiana», Banca d'Italia, *Contributi all'analisi economica*, n. 1, 1985.

[6] —— ——: «Nuove forme di accumulazione nell'industria italiana», Banca d'Italia, *Temi di discussione*, n. 52, 1985.

[7] BODO G. - VISCO I. (eds.): «Costi e profitti nel settore industriale: aggiornamenti e revisioni metodologiche», Banca d'Italia, Supplemento al *Bollettino*, n. 7, 1983.

[8] BRUNO M.: «Aggregate Supply and Demand Factors in OECD Unemployment: an Update», NBER, *Working Paper*, n. 1696, 1985.

[9] BRUNO M. - SACHS I: *Economics of Worldwide Stagflation*, Harvard University Press, Cambridge (Mass.), 1985.

[10] CIPOLLETTA L. - CALCAGNINI G. - HEIMLER A.: «Ristrutturazione ed adattamento dell'industria italiana, *Economia italiana*, n. 3, 1987.

[11] CER-IRS: *Quale strategia per l'industria*, Bologna, il Mulino, 1986.

[12] CHAN-LEE J.H. - SUTCH H.: «L'industria italiana negli ultimi 15 anni: il ciclo delle interpretazioni», in CIPOLLETTA I. (ed.): *Struttura industriale e politiche macroeconomiche in Italia*, Bologna, il Mulino, 1986.

[13] DORNBUSCH R.: «Credibility, Debt and Unemployment: Ireland's Failed Stabilization», *Economic Policy*, n. 8, 1989.

[14] GIAVAZZI F. «The Exchange Rate Question in Europe», CEPR, *Discussion Paper*, n. 298, 1988.

[15] GIAVAZZI F. - GIOVANNINI A.: «Can the Ems be Exported? Lessons from Ten Years of Monetary Policy Coordination in Europe», CEPR, *Discussion Paper*, n. 285, 1988.

[16] —— ——: *Limiting Exchange Rate Flexibility: the European Monetary System*, Cambridge (Mass.), Mit press, 1989.

[17] GIAVAZZI F. - PAGANO M.: «The Advantages of Tying One's Hands: Ems Discipline and Central Bank Credibility», *European Economic Review*, n. 5, 1988.

[18] GIAVAZZI F. - SPAVENTA L.: *High Public Debt: The Italian Experience*, Cambridge, Cambridge University Press, 1988.

[19] GRAZIANI A. - MELONI E.: «Inflazione e fluttuazione della lira», in NARDOZZI G. (ed.): *I difficili anni '70*, Milano, Etas libri, 1980.

[20] GRESSANI D. - GUISON L. - VISCO L.: «Disinflation in Italy: an Analysis with the Econometric Model of the Banca d'Italia, *Journal of Policy Modelling*, n. 10, 1988, pp. 163-203.

[21] HEIMLER A. - MILANA C.: *Prezzi relativi, ristrutturazione e produttività: le trasformazioni dell'industria italiana*, Bologna, il Mulino, 1984.

[22] MINISTERO DEL TESORO: «Ricchezza finanziaria, debito pubblico e politica monetaria», *Rapporto della Commissione di studio nominata dal Ministro del Tesoro*, Roma, 1987.
[23] ——: (Direzione generale del debito pubblico): *Il debito pubblico in Italia: 1961-87*, Roma, 1988.
[24] MODIGLIANI F. - PADOA SCHIOPPA T.: «La politica economica in una economia con salari indicizzati al 100% o più», *Moneta e credito*, 1977.
[25] OECD: *Economic Survey of Italy*, 1977.
[26] PAGANO M.: «Monetary Policy, Capital Controls and Seigniorage in an Open Economy. Discussion of A. Drazen», in DE CECCO M. - GIOVANNINI A. (eds): *European Central Bank? Perspectives on Monetary Unification After Ten Years of the Ems*, Cambridge, Cambridge University Press, 1989.
[27] SACHS J. - WYPLOSZ C.: «The Economic Consequences of President Mitterand», *Economic Policy*, n. 2, 1986.
[28] SILVANI M.: «Dipendenza energetica, struttura produttiva e composizione del commercio estero», in ONIDA E. (ed.): *Innovazione, competitività e vincolo energetico*, Bologna, il Mulino, 1985.
[29] SPAVENTA L.: «Two Letters of Intent. External Crises and Stabilization Policies: Italy 1973-77», in WILLIAMSON J. (ed.): *Imf Conditionality*, Washington, Institute for International Economics, 1983.
[30] UNITED NATIONS ECONOMIC COMMISSION FOR EUROPE: «Wage Rigidity in Western Europe and North America», *Economic Survey of Europe: 1987-88*, 1988, pp. 100-10.

Restructuring Processes, Technological Change and Economic Growth (°)

Innocenzo Cipolletta - Alberto Heimler (*)
Confindustria, Roma

1. - Introduction

In recent years, various attempts have been made to provide an explanation for the restructuring processes which characterised the Italian economy in the last decade. As the data available gradually increased, so the empirical analyses became numerous and broader and more articulated. Setting aside some simplistic interpretations which tend to ascribe the restructuring to a single cause (exchange-rate policies) or to a single player (entrepreneurs, trade unions, the government), a number of articulated models appeared in the literature.

We refer in particular to the role of relative prices in guiding the allocation of resources and to their importance, together with technical progress, in generating restructuring processes (Heimler and Milana [13]). Furthermore, short-term macroeconomic policies can be reinterpreted as a factor for "involuntary" restructuring (Cipolletta [5]), especially if analysed together with wage, labour (Dal Co [10]), and industrial policies (Momigliano [14]). Other authors have considered as essential the flexible specialisation model as a reaction

---

(°) An extended version of this article was published in *Rivista di politica economica*, Roma, Sipi, July-August 1989.
 Advise: the numbers in square brackets refer to the Bibliography in the appendix.
 (*) The author is now working at the Italian Antitrust Authority.

against institutional rigidities (Piore and Sabel [17]); the growing specialisation of Italian industry in sectorial 'niches' characterised by reduced foreign competition (Onida [16]); and the existence of a regional pattern of industrialisation (Fuà and Zacchia [11]). More recently, a number of interpretations have assigned a special role to economic policy and to the behaviour of social partners. These include analyses of the effects of state subsidies for social security contributions and of fiscal drag as factors for loosening the exchange-rate constraints (Giavazzi and Spaventa [12]) or the hypothesis of an "imperfect deal" between industry, trade unions and the government, concluded with the objective of making the labour market more flexible and simultaneously encouraging restructuring processes (Barca and Magnani [1]): the deal, it is claimed, turned out to be imperfect because the flexibility agreed to was not repaid by a similar joint agreement as regards the restructuring processes.

The analyses of the research department of the Confederation of Italian Industry (Confindustria, [6], [7], [8] and [9]) placed the restructuring processes at centre stage and, while not denying the effects of the economic policy tools adopted (devaluation of the lira, state subsidies for social security contributions and, more generally, fiscal policy) and of the changed institutional background (labour market flexibility and the role of the Cassa integrazione guadagni (1)), sought to demonstrate that the economic policies adopted by themselves would not have been sufficient. In fact, these policies were adopted in a context in which there was a rather sustained process of technological progresses, in the absence of which it would not be possible to explain the results achieved.

In this study, which takes up and develops some of the considerations contained in the *XI Rapporto del Csc* (Confindustria [9]), we intend to deal in greater depth with the links existing between economic conditions and technical progress and also to demonstrate the importance of technological progress in moderating increases in production costs and, more generally, in promoting economic growth. We will then assess the impact of the state subsidies for social security contributions as a factor for moderating the cost of labour

---

(1) The Cassa integrazione guadagni is a sort of unemployment subsidy.

and discuss the difficulties of evaluating the degree of capacity utilisation in a rapidly changing technological background. We will conclude with some brief considerations on comparisons and on the requirements of economic and industrial policies.

2. - R&D Expenditure and Technological Progress

In the last decade, in all European countries the growth rate for R&D expenditure has been higher, even though only slightly, than that for GDP. In the United States and Japan, the increase in research spending has been much more sustained (Table 1). Research spending is a very important source of growth and competitiveness and industrial countries, even though not to the same degree, have recognised this. However, even without considering its public good nature, there is no guarantee that the amount of R&D investment actually undertaken is socially optimum.

First of all, R&D spending on a certain project has a minimum threshold which is not always in line with the small dimension of firms in some industries and in some countries. A research programme is a long-term investment characterised by very uncertain returns and, therefore, by a great uncertainty over the amount of money to be invested. As a result, small-sized companies are not only unable to raise an adequate amount of resources, but they are also unable to set in motion a research circuit with, while giving rise to long-term results, permits continual spin-offs in terms of innovations and hence regular amortisation of the research costs.

From the cyclical point of view, R&D spending needs the same external conditions which favour long-term investments: namely, adequate financial resources which are not easily available in many countries.

European countries generally have an industrial structure made up of relatively small firms (Italy is an extreme case amongst the European countries) with little integration among themselves (vis-à-vis the Japanese case) which makes a consistent research effort more difficult. Furthermore, from a cyclical point of view, the high real cost of money (above the growth rate of the economy) has the effect of

TABLE 1

R&D EXPENDITURE IN THE MAJOR INDUSTRIALISED COUNTRIES
(as % of GDP)

Country	1977-1978	1981-1982	1983-1984	1985-1986	1987
Italy	0.9	1.0	1.1	1.4	1.5
W Germany	2.1	2.4	2.5	2.7	2.7
France	1.8	2.0	2.3	2.3	2.3
United Kingdom	2.1	2.4	2.3	2.2	2.4
United States	2.1	2.4	2.7	2.8	2.8
Japan	2.3	2.4	2.7	2.8	2.8

Source: CONFINDUSTRIA [9].

redirecting investment to short-term uses, thus in practice discouraging expenditure on scientific research. This is why, again without considering the public good nature of research spending, economic policies play a decisive role in all countries, in particular with regard to basic research.

In all countries, the government finances significant amounts of research (Table 2), both directly and also indirectly (fiscal policy, state demand policies, etc.). In fact, the importance of state intervention in encouraging research spending cannot be limited solely to the financial aspects. Government demand, especially when directed to high technology products, stimulates a high level of scientific research. This channel, which is very important in promoting technological development, is very difficult to measure empirically: while its significance is probably considerable in the United States it remains limited in Europe and particularly so in Italy.

The importance of scientific research is confirmed by the results of a number of studies on the experiences of the main industrialised countries. They show that long-term economic growth strongly depends on R&D expenditure. Nonetheless, this relation is not necessarily always valid in the medium- and short-run. In fact, even countries (such as Italy) characterised by a low level of spending on scientific research can innovate by importing technology and thus attain high rates of technological progress and of economic growth. Yet such a strategy, while extremely important in the short-run, in no

TABLE 2

DISTRIBUTION OF BASIC RESEARCH IN 1983 BY SECTOR
OF ORIGIN OF SPENDING
(in %)

Country	Industry A	Government B	University C	Private non-profit making D	Non-competitive sector B+C+D
Italy	5.4	30.0	64.4	—	94.6
W Germany	17.1	23.9	58.4	0.6	82.9
France (1979)	8.9	21.7	66.8	2.6	91.1
UK (1979)	14.0	24.0	60.0	2.0	86.0
United States	19.7	15.3	56.4	8.6	80.3
Japan	29.0	10.0	58.5	2.5	71.0

Source: OECD [15].

way guaranteees that the economic growth rates achieved will be maintained in the future.

To assess the importance that technological progress has had in determining growth in recent years, the growth rate of the aggregate "private-sector economy" of the six major industrial countries has been broken down into its principal components: growth rate of the quantities of the production factors utilised in the production process (labour, capital, imports) and technical progress. To do so, it is assumed an aggregate production function:

$$(1) \qquad Y = f(\bar{K}, L, M, t)$$

where:

Y = aggregate supply (2);
$\bar{K}$ = capital *stock* (assumed fixed in the short-term);
L = number of employees;
M = total imports;
t = index of the level of technological progress

---

(2) Defined as the sum of the private-sector value-added and imports (understood as imports of intermediate goods).

the total differential of which is:

$$(2) \qquad dY = \frac{\partial f}{\partial K} dK + \frac{\partial f}{\partial L} dL + \frac{\partial f}{\partial M} dM + \frac{\partial f}{\partial t} dt$$

Dividing *(2)* by Y and assuming that firms maximise profits (3) we obtain:

$$(3) \qquad \frac{dY}{Y} = s_k \frac{dK}{K} + s_l \frac{dL}{L} \; s_m \frac{dM}{M} + \frac{(\alpha Y / \alpha t)}{Y}$$

where:

s_k = ratio between capital costs assessed *ex-ante* and total production at current prices;

s_l = ratio between labour costs and total production at current prices;

s_m = ratio between costs for imported semi-finished products and total production at current prices;

and the final term represents the increase in production due to technological progress (total factor productivity). Equation *(3)* is valid only in continuous time; it has to be transformed to the discrete case. A convenient transformation is Tornquist's approximation which has been applied to the aggregate data for private-sector production of the six major industrialised countries.

The analysis of the results given in Table 3 reveals important differentiations. The Italian economy is the only one in which total productivity provides the main contribution to economic growth for most of the years. For 5 of the 8 years considered, the growth rate of total productivity in Italy has exceeded the contributions to growth deriving from the increase of the quantitative levels of the production factors. Only Japan, where however the increase in capital *stock* is the

---

(3) It is assumed that the selling prices are exactly equal to marginal costs. As a cosnequence $s_k + s_m + s_l \gtreqless 1$ if p_k (the user cost of the *ex-ante* capital) $\gtreqless z_k$ (the user cost of *ex-post* capital). For further details see BERNDT - FUSS [3] and more recently HEIMLER [2].

TABLE 3

THE ACCOUNTING OF PRODUCTION GROWTH

Country	Year	Production	Capital	Labour	Intermediate *inputs*	Total factor productivity

(average annual % rates of change)

Country	Year	Production	Capital	Labour	Intermediate inputs	Total factor productivity
Italy	1981	0.07	0.76	−0.25	−0.87	0.44
	1982	−0.08	0.59	0.13	−0.16	−0.64
	1983	4.73	0.50	0.21	−0.35	4.36
	1984	−5.04	0.52	0.03	2.19	2.24
	1985	2.30	0.54	0.34	1.00	0.40
	1986	2.84	0.55	0.40	0.89	0.97
	1987	4.55	0.62	0.24	1.74	1.87
	1988	5.15	0.69	0.68	1.28	2.42
W Germany	1981	−0.80	0.96	−0.70	−0.71	−0.33
	1982	−1.03	0.83	−1.13	−0.36	−0.37
	1983	1.62	0.89	−0.94	0.44	1.24
	1984	3.66	0.82	0.04	1.35	1.40
	1985	2.33	1.55	0.26	0.84	−0.33
	1986	2.54	1.53	0.46	0.75	−0.21
	1987	2.50	1.53	0.27	0.91	−0.22
	1988	4.17	1.54	0.63	1.35	0.59
France	1981	0.39	1.18	−0.51	−0.44	0.17
	1982	2.66	1.07	−0.13	0.55	1.15
	1983	0.03	0.97	−0.48	−0.57	0.12
	1984	1.75	0.87	−0.70	0.56	1.02
	1985	2.13	1.67	−0.19	0.91	−0.27
	1986	3.15	1.69	0.13	1.44	−0.14
	1987	3.20	1.72	0.18	1.21	0.06
	1988	4.00	1.73	0.23	1.28	0.70

TABLE 3 continued

Country	Year	Production	Capital	Labour	Intermediate inputs	Total factor productivity

(average annual % rates of change)

Country	Year	Production	Capital	Labour	Intermediate inputs	Total factor productivity
United Kingdom	1981	−1.82	0.79	−1.99	−0.63	0.02
	1982	2.79	0.84	−1.07	1.09	1.93
	1983	4.52	0.87	−0.28	1.43	2.45
	1984	3.80	0.90	1.14	2.32	−0.59
	1985	3.28	1.25	0.60	0.65	0.74
	1986	4.15	1.25	0.31	1.57	0.96
	1987	5.18	1.26	0.90	1.72	1.22
	1988	6.28	1.32	0.75	2.97	1.10
United States	1981	2.25	1.21	0.85	0.22	−0.05
	1982	−2.68	0.87	−0.51	−0.24	−2.79
	1983	5.06	0.82	0.81	1.12	2.22
	1984	9.44	1.16	2.65	2.13	3.20
	1985	3.90	1.31	1.02	0.53	0.99
	1986	3.96	3.05	1.23	0.98	−1.32
	1987	3.96	2.82	1.40	0.68	−0.97
	1988	4.32	2.69	1.23	0.66	−0.30
Japan	1981	3.53	1.77	0.58	0.23	0.91
	1982	2.61	1.57	0.66	0.06	0.31
	1983	2.50	1.55	1.03	−0.35	0.25
	1984	6.04	1.68	0.32	1.25	2.68
	1985	4.35	3.28	0.42	−0.05	0.66
	1986	2.73	4.00	0.52	0.42	−2.14
	1987	4.48	3.82	0.63	0.46	−0.45
	1988	7.61	3.71	1.39	1.52	−0.80

Source: CONFINDUSTRIA [9].

most important growth factor, shows total productivity growth rates similar, even though slightly lower, to that of Italy. In the other European countries and in the United States, on the other hand, the contributions to growth deriving from overall efficiency improvement are much lower and in some cases even negative.

The results obtained show the importance of technical change for explaining Italy's recent economic performance. Since they were reached with one of the lowest spending on scientific research in Europe they show the importance of the "spontaneous" process of industrial reorganisation which has characterised the Italian economy over the last ten years, exploiting the growing availability of technological innovation (4).

Nonetheless, the sustained growth of Italy's total productivity hides some difficulties to adjustment due to the lack of consistent research activity which would hamper the prospects for further gains in global productivity in the future. Capital *stock* has in fact shown the most modest increases among all the countries considered (less than 1% annual growth) (5).

Capital *stock* has grown strongly in other countries, above all in Japan and the United States. These results depend on the differing formulation of monetary and fiscal policies in the different countries. In Europe, and in Italy in particular, firms have been more affected by the adverse effects of high real interest rates (Graph 1) and, as a consequence, capital *stock* growth has been relatively less sustained.

3. - Production Restructuring, Production Costs and Prices

The change of factor intensities due to technology changes not only encourages production growth, but allows for a significant reduction in production costs. In order to evaluate this phenomenon,

---

(4) A recent survey of companies' patent activities carried out by CER [4] clearly shows that even when companies carry out scientific research the patents produced are generally incremental vis-à-vis the state of knowledge; this result should be even more valid when there is no scientific research.
(5) For some recent, more detailed considerations on Italian industry's structural changes see CONFINDUSTRIA [9].

GRAPH 1

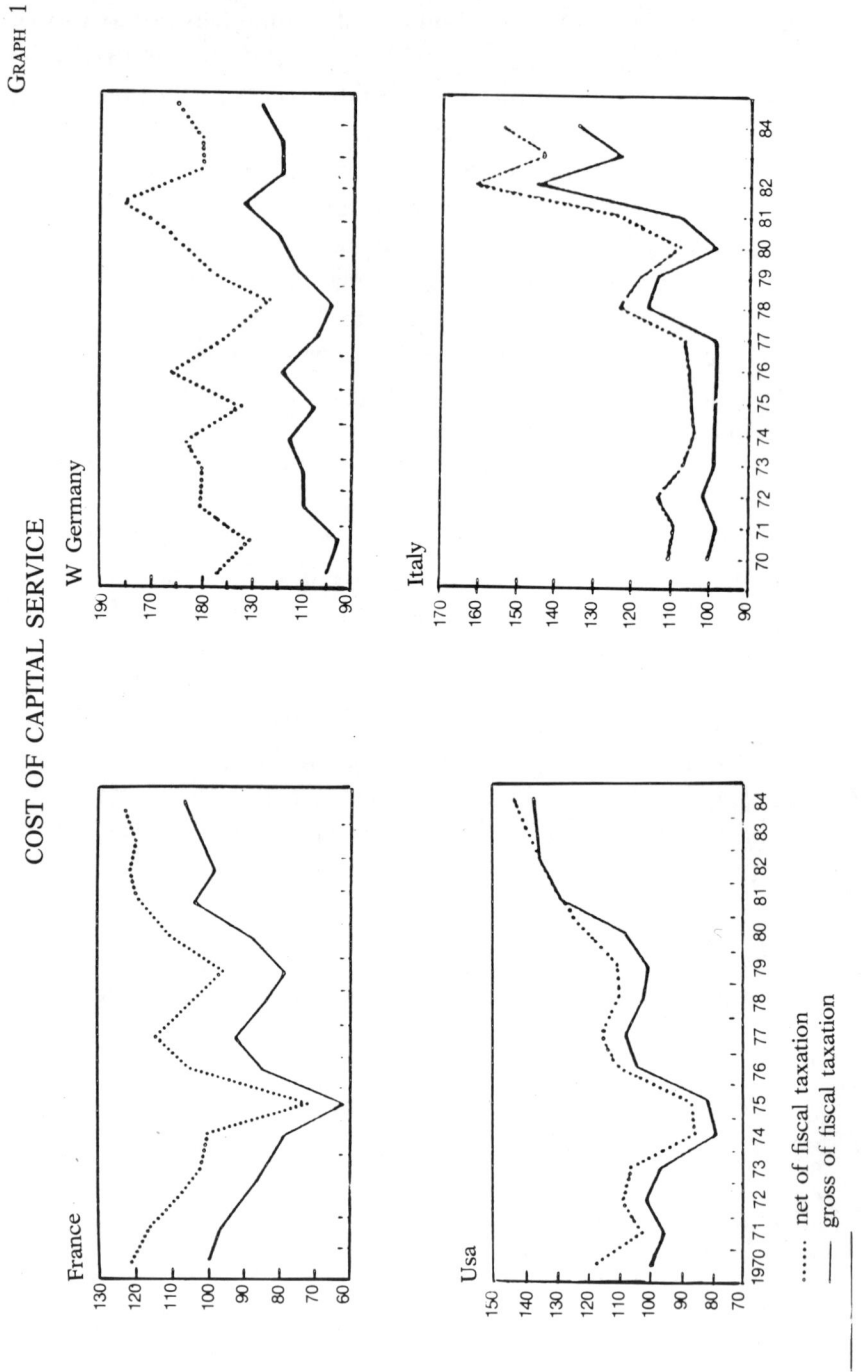

COST OF CAPITAL SERVICE

W Germany

Italy

France

Usa

.... net of fiscal taxation
——— gross of fiscal taxation

Source: Fitoussi J.P. - Phelps E.S.: *The Slump in Europe*, Oxford, Basil Blackwell 1988, for Italy, calculation from Istat and Ministero delle Finanze data.

production costs with zero change in technology have been reconstructed. When compared with the actual costs, these theoretical costs allow one to asses the extent of the restructuring processes in the various countries. The index of industrial restructuring is obtained from the ratio between effective costs and simulated costs with constant technology:

$$(4) \qquad R_t = 1 - \frac{\Sigma\, a_{it} w_{it}}{\Sigma\, a_{i0}\, w_{it}}$$

where:

a_{it} = unit requirement of the production factor i at time t
w_{it} = price of the production factor i at time t;
a_{i0} = unit requirement of the production factor i at time 0

Table 4 shows the results of the breakdown of the variations of costs for the aggregate "private sector" of the six major industrialised countries. In all countries, one can note a significant restructuring process over the period 1980-1988; this process concerned primarily Italy and Japan, where total costs, because of technical change, fell by respectively 10.5% and 9.6%. Even in the United Kingdom, the effects of restructuring were important since they allowed costs to be reduced by 8.5% while production rationalisation measures in France, Germany and in the United States produced more modest results (6).

These results show the important role that technological transformation processes played in maintaining production competitiveness. The big increase in selling prices that would have taken place without such processes would not in fact have permitted many countries, Italy in particular, to successfully face international competition and to maintain their production and employment base other than via much stronger currency devaluations.

---

(6) In any case, it should be borne in mind that the assessment of the cost reductions due to production restructuring measures has been calculated on the basis of the observed prices of production factors and therefore does not take into account the further increase which costs would show as a result of the working of the mechanism which transmits inflation to the factor prices.

TABLE 4

EFFECTS OF RESTRUCTURING
ON PRIVATE-SECTOR UNIT COSTS: 1980-1988

Country	% change in unit cost vs 1980
Italy	− 10.5
W Germany	− 2.2
France	− 3.4
United Kingdom	− 8.5
United States	− 2.5
Japan	− 9.6

Source: CONFINDUSTRIA [9].

Nertheless, higher increases in efficiency do not necesarily imply a greater ability to bring inflation under control. In fact, restructuring activities, especially in the presence of sustained rises in factor prices, do not eliminate increases in production costs; above all they do not block the cost increase mechanisms which necessitate new and continuous efficiency gains to be counteracted. If taken to the extreme consequences, such a mechanism leads to a process of deindustrialisation, or if it is interrupted due to the impossibility of continuing on the path of improving efficiency, immediately retriggers the inflation-devaluation spiral. Thus the divergences between the growth rates of the different countries' factor prices and the cost of labour in particular inevitably affect inflationary differentials (Table 5).

4. - State Subsidies for Social Security Contributions and the Restructuring Processes

The reduction of the inflationary process has been assisted not only by production reorganisation, but also by specific measures of economic policy. These have first of all slowed down the growth rate of labour costs by reducing, in 1978, the rate of social security contributions. Then, starting from 1980, by maintaining the exchange rate of the lira within the EMS relatively stable (accepting devalu-

TABLE 5

PRICES OF FACTOR *INPUT* AND OF PRODUCTION
(annual % change)

Country	Years	Capital	Labour	Intermediate inputs	Aggregate output
Italy............	1981-1983	52.7	17.9	15.2	14.3
	1984-1988	3.4	9.3	1.3	6.4
	1984	− 1.3	12.0	11.1	11.2
	1985	−10.5	10.4	7.3	9.2
	1986	26.2	7.2	−15.2	2.7
	1987	3.0	8.4	− 0.5	4.5
	1988	− 0.5	8.4	3.9	4.6
W Germany	1981-1983	16.2	5.0	5.0	4.3
	1984-1988	− 0.6	3.3	− 1.9	1.2
	1984	− 0.5	3.8	4.7	2.7
	1985	− 2.7	3.3	2.1	2.2
	1986	− 4.5	4.0	−11.8	− 0.6
	1987	2.4	3.2	− 4.7	0.5
	1988	2.2	2.4	0.4	1.5
France	1981-1983	28.6	13.0	13.3	11.1
	1984-1988	3.0	5.3	0.5	3.9
	1984	9.8	8.1	10.0	7.8
	1985	4.5	6.7	2.2	5.4
	1986	−10.4	4.2	−12.8	1.1
	1987	25.0	3.2	0.2	2.0
	1988	−13.8	4.4	3.1	2.9
United Kingdom	1981-1983	48.2	10.0	7.6	7.8
	1984-1988	− 2.8	6.3	2.2	4.3
	1984	− 1.0	4.3	8.7	5.8
	1985	− 4.4	6.5	3.9	5.2
	1986	5.6	6.9	− 3.8	1.7
	1987	− 2.4	5.6	2.8	4.3
	1988	−11.9	8.0	− 0.7	4.4
United States ..	1981-1983	25.3	6.5	− 0.2	5.7
	1984-1988	− 2.8	4.2	0.4	2.7
	1984	1.5	4.7	− 1.0	2.7
	1985	− 8.0	2.7	− 2.4	2.1
	1986	−26.0	3.5	− 2.2	2.0
	1987	9.2	4.3	5.7	3.4
	1988	9.2	5.9	1.9	3.2
Japan	1981-1983	19.1	4.5	− 0.5	1.5
	1984-1988	− 5.7	3.9	9.0	− 0.5
	1984	− 9.3	3.9	− 3.0	0.4
	1985	− 2.6	3.9	− 3.2	0.7
	1986	− 5.5	4.0	−33.3	− 2.2
	1987	− 7.4	3.0	− 4.3	− 0.6
	1988	− 3.5	4.7	− 1.2	− 0.6

Source: CONFINDUSTRIA [9].

ations which recovered *ex post* the loss of competitiveness and not also *ex ante* as had been the case in previous years). Again, starting in 1980 the reduction of the rate of social security contributions was interrupted and partly reversed; at the same time the level and the frequency of the adjustment of wages to changes in price was substantially reduced.

The reduction of the rate of social security contributions therefore encouraged a lowering of labour costs at the end of the 1970s, thus allowing domestic industry to be more competitive and by this means promoting the start of the restructuring process, helped contain the pressures acting on final prices. Nevertheless, in the longer-term, its impact on the reduction of labour costs and more in general on the reduction of the inflation rate was modest if compared to that exercised by technological change.

It is possible to measure these phenomena by assuming that the rate of social security contributions would not have been reduced: in such a case labour costs per worker would have been:

$$(5) \qquad \hat{w}_t = w_t (1 + F_t)$$

where:
$\hat{w}$ = hypothetical labour costs per worker;
w_t = actual labour costs per worker;
F_t = rate of reduction of social security contributions in year t

At this point, it is possible to reconstruct the cost of labour which would theoretically be recorded in the economy if the rate of social security contributions had not been reduced. It is assumed that without the reduction in the rate of social security contributions, labour costs would have grown at the same rate as $\hat{w}$. Nonetheless, the changes in the rates are not accrued year after year (as is the case, for example, with productivity increases), but labour costs would vary only with changes in the reduction of the rate of social security contributions:

$$(6) \qquad \bar{w}_t = \bar{w}_{t-1}(\hat{w}_t/\hat{w}_{t-1})(1 + F_t - F_{t-1})$$

where:
w_t = theoretical cost of labour in year t without the reduction in the rate of social security contributions.

Comparison of this theoretical labour cost with those that actually occurred provides an indication of the importance of these reductions in promoting a reduction in labour costs:

(7) $$Rf_t = 1 - (w_t/\bar{w}_t)$$

where:
Rf_t = the effect of the reduction in the rate of social security contributions on labour income in the year t.

Nonetheless, in addition to this reduction in the rate of social security contributions, the reduction of the importance of labour costs in total costs can also be attributed to the process of restructuring, defined as both the substitution of labour with other factors of production (capital and imports) and the "pure" technical progress. Indicating with:

(8) $$Rld_t = w_t(L_0/Y_0)Y_t$$

where:
L_0 = employment in year 0;
Y_0 = production in year 0;
Y_t = production in year t;

the labour income that would have been observed each year if labour productivity remained at the level of the base year 1978, it is possible to measure the impact of restructuring of production processes on labour costs:

(9) $$Rt_t = 1 - (Rld_t/R\tilde{l}d_t)$$

where:
Rt_t = the effect of restructuring on labour income in the year t
Rld_t = labour income in year t.

Table 6 shows the cumulate values of *Rf*, *Rt* and the ratio between the two for all the years from 1978 to 1988. It can be seen that only in the early years is the impact of the reduction of social security contributions on labour costs equal to that determined by the growth of labour productivity. However, in the more recent years the reduction of social security contributions becomes much less important (first column of Table 7) and in 1987 it is nine times and in 1988 sixteen times less important for explaining the reduction of labour costs than the increase in labour productivity.

TABLE 6

INCREASES IN LABOUR COSTS,
TECHNICAL PROGRESS AND THE REDUCTION
OF SOCIAL SECURITY CONTRIBUTIONS

Year	F (*) %	Rt (**) %	Rf (***) %	Rt/Rf
1977	1.3	2.1	1.3	1.6
1978	1.5	6.4	1.7	3.7
1979	1.4	13.3	1.5	8.9
1980	2.2	15.9	3.1	5.1
1981	3.1	16.5	4.9	3.4
1982	3.4	16.0	5.5	2.9
1983	3.3	16.3	5.4	3.0
1984	3.3	21.5	5.3	4.1
1985	3.0	24.3	4.8	5.0
1986	2.7	27.0	4.1	6.6
1987	2.4	32.6	3.5	9.2
1988	2.3	37.5	2.3	16.4

(*) Rate of reduction of social security contributions.
(**) See formula *(9)*.
(***) See formula *(7)*.

5. - Profit Margins, Investments and Plant Capacity Utilisation

In these past years when the rate of inflation strongly decreased, profit margins increased in all countries (Table 7). Nevertheless, an analysis of these gross profit margins does not allow one to check the

TABLE 7

PROFIT MARGINS ON VARIABLE COSTS
IN THE PRIVATE SECTOR
(ratio between prices and unit variable costs)

Years	Italy	Germany	France	UK	USA	Japan
1980-1983	1.34	1.44	1.40	1.48	1.49	1.34
1984-1988	1.39	1.49	1.47	1.50	1.53	1.39
1984	1.36	1.46	1.42	1.48	1.52	1.36
1985	1.36	1.47	1.44	1.49	1.54	1.39
1986	1.41	1.49	1.48	1.49	1.54	1.40
1987	1.41	1.50	1.50	1.51	1.53	1.40
1988	1.41	1.52	1.51	1.51	1.53	1.38

Source: CONFINDUSTRIA [9].

"normality" of these profits. To do that, it is necessary to compare the selling prices with total unit costs, i.e., comprehensive of the contribution of capital services.

When *extra*-profit margins, calculated as the ratio between prices and unit costs, are equal to one prices are equal to total unit production costs and *extra*-profits equal to zero. Otherwise, when *extra*-profit margins are greater (lesser) than one it means that costs are lower (higher) than prices and that producers are obtaining returns on capital which are higher (lower) than the equilibrium returns.

If profit margins are correctly calculated on total costs, profit levels are considerably scaled down for all countries (Table 8). In general, in Europe *extra*-profits were less than zero during the early 1980s. This does not mean that producers made losses, but simply that if they had invested their resources in financial markets they would have obtained higher returns. Nonetheless, in more recent years and above all in 1988, prices were labour equal to total unit costs, which implies that the return from production activities was in line with that obtainable on the financial markets (7). On the other

---

(7) If capital services, here assessed ignoring taxation, had been constructed considering the impact of taxes, they would have been higher (Graph 1) and hence the *extra*-profit margins would have been significantly lower than one.

TABLE 8

PROFIT MARGINS ON TOTAL COSTS
IN THE PRIVATE SECTOR
(ratio between prices and unit variable costs)

Years	Italy	Germany	France	UK	USA	Japan
1980-1983	0.92	0.95	0.98	0.86	1.03	1.08
1984-1988	0.97	0.99	0.98	0.96	1.08	1.10
1984	0.94	0.97	0.97	0.91	1.01	1.11
1985	0.99	0.98	0.97	0.95	1.05	1.11
1986	0.96	1.00	1.02	0.93	1.14	1.10
1987	0.97	0.99	0.95	0.97	1.11	1.10
1988	0.99	0.99	1.01	1.03	1.07	1.08

Source: CONFINDUSTRIA [9].

hand, in Japan and the United States, selling prices during the entire decade were always higher than total unit costs and investment in production activities was systematically more remunerative than financial investment.

Profit margins were influenced in a variable manner by the movements of real interest rates. An increase in the latter leads to a corresponding increase in the user cost of capital and hence, in the short-term (when the capital *stock* is fixed), an increase in total unit costs. In this case, firms are forced to seek to increase gross profit margins (maintaining the *extra*-profit margins close to one) to obtain returns in line with financial-market returns. However, the existence of competition, and also, in an "open" economy, the movements of the real exchange rate, are an effective constraint on price increases. Therefore in the early 1980s, unlike the United States which had to reduce the incidence of profits on production costs in order to remain competitive, those European countries whose currencies were revaluing vis-à-vis the dollar had the possibility to increase their gross profit margins. In more recent years, however, when European currencies were devaluing against the dollar, gross profit margins have been squeezed in Europe.

Nevertheless, in addition to the competition mechanism, other forces tend to reduce (even in the short term) the impact of high real interest rates on total unit costs. In particular, when the cost of capital rises, firms have an incentive to increase the degree of capital utilisation, in order to reduce total unit costs. This strategy has probably helped European companies to maintain the profitability of their production in line with financial returns, even when exchange-rates were relatively stable.

The relation between real interest rates and investment demand should be defined in further detail. In the 1980s, despite the presence of historically high real interest rates, a very sustained cycle of investment growth can be observed. The investment rate, measured by the share of fixed investments and GDP, rose in all countries in the second half of the 1980s to the levels which prevailed before the 1973 oil crisis. This phenomenon is particularly visible in the data expressed in constant prices. In this last decade in fact, the prices of investment goods have increased at rates which are much lower than those of the implicit GDP deflator due to the strong technological progress which has characterised them (8).

In the production of many investment goods, technological progress has so outweighed the reduction of input costs because of quality improvements that the selling prices of many products (above all, but not only, electronic goods) has fallen, in some cases even quite considerably (9). The rapid reduction of the prices of many capital goods has been passed on to the capital service cost (even though only in part due to the capital-account losses sustained by the owners of machinery, the price of which diminished and the value of which fell as the result of accelerated technical obsolescence). Reduced selling prices, falling capital use costs, and in recent years, returns from production investments in line with financial-market investments exercise positive effects on the demand for investments goods, above all in a phase of rapid technological development such as the present one, when the need to renovate plant has become a priority.

At the same time, the electronic revolution made possible by this very rapid technological progress has increasingly modified the con-

---

(8) For an analysis of these issues see CONFINDUSTRIA [9].
(9) This phenomenon is not always fully reflected in official statistics due to technical dificulties in measuring output at constant quality.

cept of flexibility of industrial systems, placing machinery at the centre stage of companies' restructuring strategies. In this situation, the level of plant utilisation is no longer solely a technological variable, but is also closely dependent on the economic conditions in which the company operates. Above all, an increase in capital intensity, as noted above, combined with high real interest rates makes a more intense utilisation of plants more advantageous.

The high degree of capital utilisation observed in recent years in all industrialised countries does not therefore represent an anomaly to be corrected, but is the result of the producers' correct reaction to the economic signals and in particular to the high real interest rates which force them to operate, as far as possible, with more than one production shift. Furthermore, the present wave of technological progress accentuates the flexibility of production and therefore represents a further stimulus to a high degree of capital utilisation.

6. - Conclusions and Some Industrial Policy Implications

The restructuring activity undertaken by Italian industry cannot be attributed to any one single cause but to a multitude of factors, in a chain of sequences in which time and the available technical progress have played an important role.

Analysis of the restructuring processes of the last decade reveals that the strong points of Italian industry were its good adaptability to demand and the pursuance of up-to-date technologies for many of its production processes. Corporate reorganisation enabled firms to improve the quality of production, save on costs and gain significant market shares also in the international market.

It is very important to continue on the path of improving the quality of production and of continuing to restructure. The increases in the efficiency of production depend both on the processes of technical progress endogenous to the firm and on the improvement of the quality of the factors of production. While endogenous technical progress is related to the shumpetarian nature of the firm, factors of production are acquired on the market. The quality and the effectiveness of the factors of production can be improved by government programs. In this regard, government intervention in the fields of education, professional training and scientific research are of particular importance. And so are government interventions in the social and economic infrastructures directed, in particular, at improving the quality of public services. An improvement in the efficiency of public services would have significant effects on the competitiveness of industrial products (Confindustria [6]). Furthermore, a program of public investment would generate income and employment and also, which is very important, create and improve the professional and entrepreneurial skills of the firms that would supply them.

It should be borne in mind that public spending arises from the need to satisfy collective needs such as, for example, national defence or public health and does not have the explicit purpose of modifying a country's industrial structures. Nevertheless, public spending policies could generate important effects on the industrial system, stimulating the growth of high technology industries and therefore indirectly modifying comparative advantages.

BIBLIOGRAPHY

[1] BARCA F. - MAGNANI M.: *L'industria tra capitale e lavoro. Piccole e grandi imprese dall'autunno caldo alla ristrutturazione*, Bologna, il Mulino, 1989.

[2] HEIMLER ALBERTO: «Productive Restructuring, Costs and Profit Margins in the Industrial Countries», *Journal of Policy Modeling*, n. 3, pp. 345-59, 1989.

[3] BERNDS ERNST R. - FUSS MELVYN: «Productivity Measurement with Adjustment in Capacity Utilization and Other Forms of Temporary Equilibrium», *Journal of Econometrics*, n. 33, 1986, pp. 7-29.

[4] CER: «L'attività innovativa in Italia: i brevetti nell'industria», *Rapporto*, n. 6, 1988.

[5] CIPOLLETTA I. (ed.): *Struttura industriale e politiche macroeconomiche in Italia*, Bologna, il Mulino, 1986.

[6] CONFINDUSTRIA: *VIII Rapporto sull'industria italiana*, Roma, Sipi, 1986.

[7] — —: «Ristrutturazione e competitività nell'economia italiana», *IX Rapporto Csc*, Roma, Sipi, 1987.

[8] — —: «Squilibri commerciali e aggiustamento produttivo nei paesi industriali», *X Rapporto Csc*, Roma, Sipi, 1988.

[9] — —: «Progresso tecnico, investimenti e politiche industriali», *XI Rapporto Csc*, Roma, Sipi, 1989.

[10] DAL CO M. (ed.): *Le leggi della politica industriale in Italia. Dalla ristrutturazione all'innovazione*, Bologna, il Mulino, 1986.

[11] FUA G. - ZACCHIA C. (eds.): *Industrializzazione senza fratture*, Bologna, il Mulino, 1983.

[12] GIAVAZZI F. - SPAVENTA L.: «Italia: gli effetti reali dell'inflazione della disinflazione», *Rivista di politica economica*, n. 7-8, 1989.

[13] HEIMLER A. - MILANA C.: *Prezzi relativi, ristrutturazione e produttività. Le trasformazioni dell'industria italian*, Bologna, il Mulino, 1984.

[14] MOMIGLIANO F. (ed.): *Le leggi della politica industriale in Italia. Dalla ristrutturazione all'innovazione*, Bologna, il Mulino, 1986.

[15] OECD: *Structural Adjustment and Economic Performance*, Paris, 1987.

[16] ONIDA F. (ed.): *Vincolo estero, struttura industriale e credito all'esportazione*, Bologna, il Mulino, 1986.

[17] PIORE M. - SABEL C.: *Le due vie dello sviluppo industriale. Produzione di massa e produzione flessibile*, Torino, Isedi, 1982.

The Role of Monetary and Financial Policies in the Restructuring of Industry [*]

Stefano Micossi - Fabrizio Traù
Centro Studi Confindustria, Roma

Introduction

This study provides an overview of the transformations which between the second half of the 1970s and the first half of the 1980s characterized the evolution of the real structure and the forms of financing of Italian companies. It places special emphasis on the role of aggregate demand management policies, monetary and exchange rate policies in particular.

Section 1 provides a brief review of the essential features which characterized the economic and financial conditions of industrial companies at the beginning of the period under study, and then proceeds to trace an overall picture of the phases and structural characteristics of the adjustment process that took place subsequently.

Section 2 discusses the role which economic policy measures (monetary policy in particular) played at the aggregate level in progressively reducing the macroeconomic imbalances inherited from the early 1970s.

In Section 3 the effects of the policies adopted are analyzed at a

---

[*] This article was published in the *Rivista di politica economica*, Roma, Sipi, July-August 1989.

disaggregate level emphasizing, in particular, the role that the size of a company played in the restructuring process.

Section 4 presents some concluding observations concerning the current situation and prospects on the analyzed phenomena.

1. - Financial Structure, Economic Policies and Production Adjustment

1.1 - At the start of the 1970s the Italian economy had an underdeveloped financial structure in which almost all financial flows were intermediated by credit institutions (for the most part banks and special credit institutions (1).

In addition, financial markets were segmented by specialization constraints that limited the activity of intermediaries (2). There was no money market (3) and the functioning of the bond market was closely controlled by the central bank, which actually "managed" the prices and quantities of new issues and intervened to stabilize prices in the secondary market (4). Finally, the state intervened with a wide range of controls, constraints, subsidies and different fiscal treatment of earned interest for aggregate and sectoral economic policy goals (5), thus aggravating market segmentation and reducing flexibility and transparency in prices and yields of financial instruments.

With regard to the final users of the resources, by the end of the previous decade the financial position of public and private sector enterprises had already considerably worsened and this adverse trend continued during the first half of the 1970s.

---

(1) On average, for the period 1971-1975, instruments issued by these institutions represented 87% of household financial savings and also provided some 77% of the economy's (households and firms) total financing; equity capital represented under 10% of total liabilities (excluding the Ministry of State Participations' endowment funds). See COTULA - MORCALDO [22] and FANNA et AL. [25].

Advise: the numbers in square brackets refer to the Bibliography in the appendix.

(2) See BISCAINI et AL. [8].
(3) See TARANTOLA [50].
(4) See BIANCHI [6].
(5) See PONTOLILLO [45].

The private sector, in particular, saw a sharp reduction in margins as a result of the 1969 wage explosion. The effects of a legislation that placed severe constraints on the use of the labor force and the growing unrest at the factory level worsened the picture (6).

The deterioration of the balance sheet accounts was the main, but not exclusive cause of a strong increase in borrowing from credit institutions. Favorable interest-rate conditions, made even more attractive by fully deducible interest payments for tax purposes and other incentives, also contributed to swelling the demand for loans, while the underdeveloped state of the market and the onerous tax system discouraged bond and share issues. Companies' high indebtedness, especially to banks, made the banking system an effective vehicle for monetary restriction (7).

The early 1970s also saw the dissipation of the conditions of financial stability which had prevailed in the previous decade, with the progressive acceleration of inflation and the emergence of a large foreign deficit. Both phenomena took on alarming dimensions after the first oil shock (October 1973). Faced with this turn of events, fiscal policy became expansionary; monetary policy reacted slowly and, by means of an increasing number of administrative measures, tried to reconcile the traditional objectives of encouraging investment with the need for restriction. Interest rates did not rise at the same pace as inflation; inflation magnified the distorting effects of widespread controls and created, at the same time, incentives for operators to find a way around them.

Thus, on the one hand, there was an attempt to limit restriction to short-term credit with the introduction of the ceiling on bank loans and simultaneously stimulate the demand for securities, to be held for compulsory-reserve purposes (8). On the other hand, the limited increase in interest rates led to the formation of a large interest-rate differential with respect to foreign markets and (in spite of significant exchange rate adjustment between 1972 and 1973) to large losses in official reserves in the attempt to support an unrealistic level for the

---

(6) For an analysis of the problems affecting the "real" side of the industrial sector in these years see BARCA - MAGNANI ([5], Chapter II).

(7) On these aspects see also MONTI et AL. [42].

(8) See FAZIO [26].

lira in foreign exchange markets. The brief experiment with the double exchange rate market (between 1972 and 1973) provided ample evidence, as shown by the large gap between official and free market rates, of the size of imbalances that had accumulated in the economy (9). Foreign currency controls were also tightened (10) to check the flow of savings abroad. These measures, inevitably, made it more difficult to finance swelling external trade and current deficits.

1.2 - The year 1975 marked the period of maximum imbalance in corporate balance sheets and profit and loss accounts. Financial indebtedness of the manufacturing companies in the Mediobanca sample, which refer mostly to large companies, exceeded 130% of sales and was strongly concentrated in short-term borrowing. National accounts data show that in the manufacturing sector gross operating margins had fallen to a minimum of 18.5% (as compared to 24.1% in the 1970-1974 period) (11). In particular, conditions in large companies were most worrisome (see Riva - Silva [46] on this matter), so much so that some industry experts suggested transforming their debts to banks into shares as a first step in a general rescue operation.

In 1976, this climate of growing alarm and lack of confidence culminated in the closing of the foreign exchange market and a sharp devaluation of the lira. In the ensuing months, a stabilization programme for public finance was eventually enacted and monetary policy became severely restrictive and strengthened by administrative measures taken to control imports (the advance-deposit scheme).

These measures led the way to the correction of the real and financial imbalances that had accumulated in the Italian economy and, in particular, in company accounts.

In 1976-1977 economic policy managed a «text-book» combination of an exchange rate depreciation accompanied by a restrictive budgetary policy and a less accommodating monetary policy (12). The manoeuvre was assisted by a favourable international situation; world

---

(9) See Micossi - Rebecchini [40].
(10) See Micossi - Rossi [41].
(11) See Landi [37], Frasca - Marotta [29].
(12) See Spaventa [49], Giavazzi - Spaventa [31].

demand had begun to expand again at a sustained rate and the depreciation of the dollar helped contain the cost of imported materials and, more generally, exchange rate stability in the EMS (13).

As it is known, this aggregate manoeuvre was integrated with a series of other measures directed at improving company balance sheets (14). On the one hand, in fact, a series of financial support measures were instituted or strengthened (15). On the other hand, the state intervened directly with measures aimed at reducing social security contributions and granting various fiscal benefits as well (16). Rescue interventions to save troubled companies by the State Participations Ministry were also decisive.

In a still decidedly inflationary climate, the effects of these measures on the state budget were largely compensated for by the erosion of personal income caused by fiscal drag (17). The government budget was therefore assigned the important task of correcting the considerable shift in the distribution of income which had taken place in the first half of the decade in favour of labour.

The effectiveness of these measures aimed at modifying the price trend was further enhanced by agreements with the trade unions, stipulated in early 1977, which mitigated wage indexation and, by increasing management's power *vis-à-vis* the labour force, brought about sizeable increases in productivity in the following years (18).

With regard to balance sheet structure, the joint effect of inflation and an accomodating credit policy kept the real interest rate down to negative values, thus bringing about a remarkable reduction in company indebtedness.

---

(13) An improvement in terms of trade works as a "reverse supply shock": it contains the increase of the product wage below that of real wages (deflated with consumer proces), thus creating room for an increase in margins. See ONOFRI - SALITURO [44].

(14) See FRASCA - MAROTTA [29].

(15) See in particular Law n. 183 of 1976 in support of small and medium-sized companies and the measures aimed at promoting restructuring in production, particularly law No. 675 of 1977 and law n. 187 of 1978.

(16) In addition to completely deductible interest payments, we refer to the various laws for the monetary revaluation of assets and the norms regulating deductions for losses and accelerated amortizations.

(17) See GIAVAZZI - SPAVENTA [31].

(18) See BARCA - MAGNANI [5], Chapter II.

1.3 - Between the late 1970s and the early 1980s the Italian industrial system underwent an adjustment process that radically modified its production and financial structure.

Appropriate economic policy measures were undertaken to reserve the erosion of profit margins caused by the wage and oil shocks and allow companies to benefit from the considerable reduction in real indebtedness due to inflation.

The improvement in companies' financial position supported their efforts while they were starting to modify factor proportions (towards capital deepening) and were accelerating the process of "farming out" part of their production (a process underway since the beginning of the 1970s) in order to reduce the rigidity of factors in the production process (19). Increased financial resources and greater production flexibility triggered a profound transformation in the organization and financing of the production cycle.

In this context, technical progress also played a crucial role: in fact, it allowed the existing capital stock to adjust quickly to the great changes which the market was undergoing and provided a decisive contribution toward adjusting the production structure to a demand which was increasingly differentiated and unstable (20).

On the one hand, the capital stock was gradually "modernized", "targeted" investments brought about structural changes in the organization of the production process and the capital/output ratio underwent a contration (21). On the other hand, the technological change further accelerated the deverticalization of the industrial structure, emphasizing production specialization in the division of labour among companies.

The multiplication of market relations among companies helped contain the size of larger production units and at the same time

---

(19) With regard to the changing relations between production factors in the adjustment process see HEIMLER - MILANA [34]; on the structural characteristics of vertical "disintegration" in Italian industry see CER [13].

(20) On this point see BIANCHI [7].

(21) For a profile of investment activity in the period under study, see the analyses in LABORATORIO DI POLITICA INDUSTRIALE [36], ZANETTI [55], GROS-PIETRO - ROSA [33], BARCA - MAGNANI ([5], Chapter 3 and 4). With regard to the progressive move from an acceleration-type accumulation process to one responding instead to the need to maintain production flexibility via marginal "upgrading" investments, see also BANCA COMMERCIALE ITALIANA [2].

created the conditions for increasing the average size of smaller units. In larger units there was also deep reorganization of the entire structure of activities, with a contraction of capital spending and an increase of the range of financial activities (22). Conversely, investment activity (more oriented towards increasing capacity) accelerated in small and medium-sized companies: these also recorded relatively higher value-added growth rates (23).

During the course of the 1980s, the differences in growth rates between companies of different size appeared increasingly linked to the effects of deverticalization; "specialization" ended up by encompassing activities that lay outside the narrow production sphere, and large companies increasingly tended to focus on the co-ordination and management of the growing flux of transactions between deverticalized units (24).

As far as financial structures were concerned, the decline in capital/output ratios compounded the effects of inflation on debt by helping to reduce external financial needs.

Deverticalization of production also helped contain the financial requirements per unit of output and at the same time favoured increasing recourse to commercial credit, helping to relax the dependance on bank credit (25).

The balance sheet improvement received further impulse from better management of financial resources; keener attention to financial optimization led to lower circulating capital and strengthened the contribution of interest income to profitability (26).

Company size was important in influencing the adjustment process also on the liability side. On the one hand, borrowing was influenced by the differences in growth rates by size groups: for given self-financing, financial requirements were proportionately higher for

---

(22) See Traù [51], Frasca - Marotta [29], Siracusano - Tresoldi [48].
(23) These phenomena are specifically analyzed in Barca - Magnani ([5], Chapter 6). See, in any case, section 3 below.
(24) See in particular Zanetti [56], Centrale dei Bilanci [12], Contini [20].
(25) See on specific aspects of this phenomenon Banca Commerciale Italiana [1], Onado [43], Traù [51], Cipolletta et Al. [17], Conti [19], Frasca - Magnani [29].
(26) With regard to the characteristics of the "financing" and the logic of its development, see the exemplary analysis in Vaciago [54].

the smaller companies. On the other hand, the development of sophisticated financial management by large companies contributed to separating the evolution of large (which are able to carry out a fully-fledged in-house financial function) from those of small companies (persistently dependent on industrial margins and bank credit for their expansion (27).

2. - Macroeconomic Imbalances and Monetary Policy

2.1 - At the macroeconomic level, the decision to enter the EMS and, therefore, stabilize the exchange rate and adopt tight monetary and credit policies, was very important in changing the behaviour of the private sector and fostering adjustment.

The lira effective nominal exchange rate, that had already depreciated by almost 24% between 1972 and 1974, fell by a further 21% in 1976 and almost 16% in the following two years; this brought about an improvement in competitiveness (on the basis of wholesale prices of manufactured goods) of 8.5% between 1975 and 1978 (12.5% if the period is extended to 1972) (28). These gains appear even greater if competitiveness is measured on the basis of relative labour costs (Graph 1). There were therefore important non-neutralities, despite high wage indexation. The result was a 33% increase in exports by volume during the period 1976-1978, while imports rose by 22%; the trade balance improved from a deficit equivalent to 4% of GDP in the period 1974-1975 to almost equilibrium in 1978; the current account balance went from a deficit equivalent to 2.5% of GDP to a surplus of similar size (29).

---

(27) See again TRAÙ [51], FRASCA - MAROTTA [29], SIRACUSANO - TRESOLDI [48].
(28) See MICOSSI - REBECCHINI [40].
(29) See BANCA D'ITALIA [3]. It should be noted in this connection that increases in income tax rates of about 5% significantly squeezed households' disposable income, which, as noted earlier, was being further reduced by inflation induced fiscal drainage. GIAVAZZI - SPAVENTA [31] points out that, according to CER evaluations, the fiscal drag reduced real wage income by more than the 4.6% overall increase in the tax rate. In this context, therefore, a significant deceleration in domestic demand allowed an increase in net exports.

It should be emphasized that, while leading to sizeable increases in nominal interest rates (Graph 4), monetary policy remained accomodating, with money supply increasing faster than nominal GDP (30); interest rates and the average cost of company financing remained below inflation (31). Inflation, measured by consumer prices, remained around 17% in 1976 and 1977 and fell to 12% the following year; in the first two years, however, the increase in wholesale prices was much greater (more than 43% against approximately 36% for consumer prices) (32). As a consequence, real wages, based on consumer prices, grew more than product-wages; the change in relative prices was thus an important source of improvement in company margins. Graph 2 shows that the increase in real wages remained below that of product wages until the early 1980s (33).

The significant increases in productivity, due primarily to economic recovery, exemptions from social security contributions and de-indexation of employment severance pay contributed to this process. On the basis of national accounts data, between 1975 and 1979 exemptions from social security contributions brought about a reduction of 4% in industry's total labor cost. In the same period the increase in labour productivity totalled 22%.

Overall, profits increased considerably: the ratio between gross margins and value-added rose to above 30% (Graph 3) (34).

At the same time, the powerful mechanism of wealth transfer from creditors (households) to indebted sectors (the Treasury and companies), via inflation and negative real interest rates, also worked to restore these sectors' financial positions (35). The ratio between companies' total debt and output fell by some 15 points (Graph 5).

---

(30) See CARANZA - FAZIO [11].
(31) See MACCHIATI ([38], Chapter 1.4).
(32) See again BANCA D'ITALIA [3].
(33) The product wage is approximated by using the wholesale price index as the deflator.
(34) See on these aspects also BARBONE et AL. [4], CIPOLLETTA [16], BARCA - MAGNANI [5], GIAVAZZI - SPAVENTA [31].
(35) FRASCA - MAROTTA [28] estimate this average wealth transfer for the period 1976-1980 as approximately 4% of GDP. Lower values have been obtained by FANNA et AL. [25].

Controls on capital flows, which prevented families from looking abroad for alternative opportunities for their savings, represented the additional condition that facilitated this transfer of wealth.

2.2 - By the end of the 1970s, the Italian economy had re-established its trade balance, but still suffered from high inflation — more than twice the European average — and imbalances in government finances had not been reduced. Companies had improved their financial position but they were still lacking in efficiency. A significant acceleration in investments had already occurred in the 1979-1980 period (industry's investments had increased by almost 30%), following the improvement in profit margins of the previous two years and the strong acceleration in demand after 1978. These investments were only in part aimed at expanding capacity; their primary objective was rather to rationalize the production process and save labour (36). In these years, the return to capital still remained at modest levels, especially if company balance sheet accounts are adjusted for the effects of inflation (37).

After the new oil shock (in the fall of 1979) the macro-policy attitude of the main industrial countries changed; they adopted strict, medium-term, anti-inflationary strategies. Furthermore, the implementation of the European exchange rate agreement, in March 1979, started to constrain European currencies; Italy had adhered to the agreement, albeit with a wider currency fluctuation band (6% versus 2.5% for the other countries), recognizing that accomodating monetary policy and devaluations were liable to bring only temporary improvements to profitability and competitiveness.

During 1979 inflation started to pick up again, rising above 25% in the first quarter of 1980. Both the trade balance and the current account balance plunged back into the red. Monetary policy became very restrictive and interest rates quickly rose above inflation (38). The real exchange rate began to rise, as the Bank of Italy refused to

---

(36) This resulted, especially for large companies, in an acceleration of the plant replacement rate (see BARCA - MAGNANI [5], Chapter III).

(37) See FILIPPI [27].

(38) See BANCA COMMERCIALE [1], CARANZA - FAZIO [11], MICOSSI - ROSSI [41].

accomodate inflation differentials with nominal exchange rate depreciations (39).

By 1986, inflation had fallen to 6% and the current account was again in the black, while economic activity was again expanding after the downturn of the early 1980s.

The explanation for these results are essentially to be found in the extraordinary restructuring efforts which had pervaded the production processes, leading to exceptional increases in productivity (on average, 4% per year, with increases of close to 9% for large companies) (40).

Employment fell by about one million units in industry (and by more than 400 thousand units in agriculture). This drop was compensated for by an increase of 1,700,000 units in the service sector, mostly in public administration. The social costs of restructuring were reduced by the state's Cassa Integrazione Guadagni (the wage supplementation fund) and by considerable increases in income-support measures. The new economic policy, consisting of a strict monetary policy and a more expansionary budget policy (which was soon reflected in larger deficits), inverted, therefore, the previous decade's economic policy mix.

On the demand side, the difficulties encountered by Italian companies in the European market, following the Italian decision to join the EMS, were partially offset, for the entire first half of the 1980s, by the progressive appreciation of the dollar which sustained the competitiveness of Italian products in the North-American market (41).

The deceleration in inflation was considerably favoured by the drop in raw material prices starting in 1983, the depreciation of the dollar starting in 1985, and the oil counter shock in 1986. On the domestic front, real wage increases were more contained than in the 1970s but still approached 2% per year; the rise in unit labour cost

---

(39) See CIAMPI [14], GRESSANI et AL. [32], BARCA - MAGNANI ([5], Chapter II). The real appreciation of the exchange rate vis-à-vis the other European currencies was, in the years of the EMS, in the order of 16%; it took place primarily during the system's first years of life.
(40) See BANCA D'ITALIA ([3] 1986), BARCA - MAGNANI [5], CIPOLLETTA - HEIMLER [18].
(41) See DE CECCO [23].

was still close to that of prices and hence its contribution to slowing inflation can be considered relatively minor (42).

During the 1980s, the financial position of companies continued to strengthen. Gross operating margins continued to grow relative to value-added, and net interest payments fell (Graph 6). Return to capital went up considerably as well (43).

3. - Adjustment Paths in the Restructuring Process

3.1 - A more complex picture of the role played by monetary policy in the restructuring process can be derived by looking at the different adjustment paths that characterized the "transition" from the 1970s to the 1980s of companies of different size groups (44).

Graph 7 shows gross margins over value-added for manufacturing (from Istat's survey of gross national product for the years 1975-1985) (45).

The chart shows that in the second half of the 1970s the most significant differences in profit margins between different size classes are found during the years of the exchange rate depreciation and during the period of expansion at the end of the decade. The improvement in margins between 1975 and 1976, the years of the depreciation, seems to concern, for the most part, the larger com-

---

(42) See BANCA D'ITALIA ([3], 1986, Table b21, p. 127). The effect of expectations in slowing down inflation shouldn't be overestimated in this context. A study by GIAVAZZI - GIOVANNINI [30], in fact, shows that, in Italy, a "leap" in expectations occurred only after the 1984-1985 period, when the deceleration of inflation was already well under way.

(43) See CENTRALE DEI BILANCI [12].

(44) The lack of homogeneity which characterizes company distribution by size on a sectoral level must be kept in mind. "Large" and "small" company behaviour reflects in fact the respective sectors' development trend, which can be extremely differentiated in the period under study. With reference to specific aspects of these phenomena see, among others, LANDI [37], CANOVI [9], SEMBENELLI [47], BARCA - MAGNANI [5]).

It must also be pointed out that the heterogeneity of available data sources greatly impaired the comparison of various real and financial aspects of company behaviour.

The results which are presented here should therefore be considered only as a simple stylization of the observed phenomena.

(45) We would like to thank F. Barca of the Bank of Italy's Servizio Studi for kindly providing us with the revised Istat series on gross national product. For a discussion on the revision criteria see BARCA - MAGNANI ([5], Chapter 6, note 1).

panies, while the smaller companies appear to be the ones to record the largest increases in 1979 and 1980.

In the first case, profit margins' differentials do not appear linked to productivity (in the same period gross output per worker increased, in fact, by similar amounts in all different size classes); a larger size seems, therefore, to provide greater capacity for exploiting exchange rate depreciations (larger companies seem able to successfully maintain higher prices in export markets) (46).

In the second case, however, the faster growth in margins recorded by small and medium-sized companies is accompanied by a progressive differentiation in productivity increases, which appear to be inversely correlated to size (47).

At the same time, the divergence between "large" and "small" companies's trend of growth, which characterizes the production structure for the entire second half of the 1970s, becomes significantly marked during the 1978-1980 economic expansion. A sustained level of employment, higher rates of growth of value-added (48) and, above all, a remarkable increase (Graph 9) in investments (in a phase in which capital accumulation in medium and large companies was instead progressively slowing down) make the smaller companies' propensity to expand a structural phenomenon.

If the exchange rate policy that followed the first oil shock seems to mostly benefit larger companies, a rather "permissive" monetary policy, however, allows, throughout this period, smaller companies to expand production capacity and achieve high levels of efficiency. Thus, self-financing among smaller companies rose constantly (49) while in-

---

(46) It should be noted that behavioural differences among different size companies are not linked to sectoral differences (export prices in sectors characterized by a higher average company size do not rise faster than in other sectors); see on this point ICE [35], 1987.

(47) In the intermediate years, growth in profit margins do not show appreciable differences among different size-classes although, between 1976 and 1977, ULC in larger companies appear to have risen faster. This might indicate that the exemptions from social security contributions granted in those years (see Section 2) were substantially irrelevant when compared to the relative performance of companies in the various size classes; therefore, with regard to firm's size, the impact of fiscal policy appear to have been substantially neutral.

(48) See in particular BARCA - MAGNANI ([5], pp. 115 and 228).

(49) With regard to this "virtuous" circle between accumulation and efficiency, which the 1978-1980 recovery fostered in small companies, see again BARCA - MAGNANI ([5], Chapter 6).

debtedness among them remained unchanged (and in some cases it even increased, thanks to their sound initial financial position) (50).

Access to "external" financing was also facilitated by credit availability. The larger companies' reduction of investments caused, in fact, a drastic contraction in their borrowing (51).

In this perspective, up to the end of the 1970s, monetary policy was able to guarantee the availability of resources to the small companies while they were trying to "reposition" themselves for the new kind of relations among production units implied by the deverticalization process (52). The lack of financial constraints was also favoured by the expansion of commercial credit (a consequence of expanding market relations among companies) as both an alternative and a substitute financial instrument to bank credit; the very logic of the restructuring process implied, in fact, in this phase, a relaxation of the constraints imposed by credit intermediaries (53).

3.2 - The shift to tight exchange rate policy and the return of positive real interest rates (1979-1980) had a heavier impact on smaller companies that were more dependent on the evolution of demand and bank financing.

While at the beginning of the 1980s the contraction of medium and large companies' labour and capital inputs led to a strong recovery of productivity (54), the effects of a "strong" exchange rate

---

(50) The available data does not allow us to see the phenomenon between 1978 and 1980; it should, however, be noted that in 1984 the leverage of smaller companies was on average at the same level as in 1973; on the evolution of the indebtedness of companies with respect to their size see also TRAÙ [51].

(51) The rapid reduction of large companies' indebtedness must also be linked to the effects of inflation (see Section 2.1), the impact of which was particularly heavy in the "base" sectors (ONADO [43]), characterized by a greater number of large companies.

(52) With regard to the relation which links the differentiated growth of the small company to its changed role in the restructuring process, see also DELLI GATTI - SARACENO [24].

(53) With reference to the role played by commercial credit in the deverticalization process of the production structure, see, in particular, BANCA COMMERCIALE ITALIANA [1]; on the almost total absence of credit rationing effects, despite the application of a ceiling on bank loans, see ONADO [43].

(54) Contrary to what happens to all other size classes, value-added per employee in companies with more than 200 employees also rose (Chart 8) between 1981 and 1982. If employees in the "wage supplementation fund" (CIG) are taken into account, labor productivity in larger companies is even higher between 1980 and 1985 and, therefore, the data represented in the chart might underestimate the relative efficiency increases that in fact occurred.

(55) and the slowdown in demand, with productivity increasingly worsening, led to a strong squeeze in small and medium-sized companies' margins (56).

Rising financial burdens from higher interest rates (and persistent indebtedness) led to a significant worsening of the "leverage" for small companies and hence to lower profitability (57).

Given a negative correlation between size and growth (58), the deterioration of smaller companies' profitability in the first half of the 1980s was reflected in increasing dependence on "external" financing. The increase in borrowing was concentrated in short-term instruments. There is clear evidence for this phenomenon in the years 1982-1986, as shown by the comparison between the rate of change in size (expressed by gross assets) and short and medium-long term financial indebtedness for some 11,000 manufacturing companies from the Centrale dei Bilanci database, broken down into nine size classes (Graphs 11 and 12) (59).

The graphs show that in general, growth of assets and borrowing were correlated to company size.

---

(55) Even the US market, which in the early 1980s helped Italian companies to overcome the impact of joining the EMS, was not sufficient to prevent the export performance being different for different size companies. Although probably not exempt from the effects of sectorial composition, the Mediocredito Centrale data indicates that between 1978 and 1984 exports of manufacturing fell from 35.3 to 30.9 percent of total sales for companies with 11 to 20 employees and from 34.3 to 32.2 for those with 21 to 100 employees, while companies with more than 300 employees maintained their export quotas practically unchanged from 31.2 to 30.6 percent. The difference in performances is certainly related to the greater capability of larger companies to differentiate market outlets (in the same period the EEC market export share goes from 60.8 to 56.2 for companies with 11 to 100 employees and from 50.8 to 38.0 for companies with more than 400 employees). It is, therefore, likely that smaller companies, more dependent on the European market, also suffered a greater reduction in profit margins.

(56) Between 1982 and 1986 "unit profits", measured as the ratio between net operating margin and sales, grew at decreasing rates (to negative figures) as company size decreases (CENTRALE DEI BILANCI, [12]). On the subject of medium-large companies' strong recovery of margins from the beginning of the 1980 see also LANDI [37], CONTI [19].

(57) With regard to the different levels of profitability in small and large firms, see TRAÙ [51] and [52]. On the different impact of high interest rates on companies of different size see SIRACUSANO - TRESOLDI [48]. On the "structural differences" of interest payments per unit of debt, see TRAÙ [52], SEMBENELLI [47]. On the role of financial revenues on large and small companies' profitability, see also FRASCA - MAROTTA [29].

(58) See TRAÙ [53].

(59) Size is defined on the basis of the number of employees in the first year.

In medium and large companies, in fact, assets grew more than their debt, although at relatively lower rates than in smaller companies. Small and medium-small companies, which grew on average relatively faster, seemed to replace long-term debt (which expands less than size) with short-term debt (which expands more).

This phenomenon is probably related to the effects of changes in commercial credit policy. If, in fact, in general, the expansion of commercial credit as an instrument for regulating transactions between companies softens the impact on the economic system of restrictive credit measures, it creates, however, distortions at a disaggregate level because of companies' different contractual power (60). The fact that in the first half of the 1980s the small sized companies registered an increase in "net" commercial credit (while medium-large companies tended to reduce it) (61) could lead to the conclusion that growth of their short-term liabilities may reflect rising difficulty in maintaining control over the terms of the credit granted or received.

In this perspective, the changes in firm's external financing contributed to reinforce the differentiated effects of monetary policy (a "transfer" of the burden of credit restrictions on smaller companies).

4. - Financial Structure and Monetary Policy: The Inheritance of a Decade of Transformations

The restructuring process and monetary and exchange rate policies which were implemented during the 1980s have had profound consequences on the financial structure of the Italian economy, and hence on how monetary policy is used and on the efficacy of its action.

Also thanks to specific interventions to support financial markets (62), the adoption of restrictive policies has led, in fact, to a more

---

(60) See TRAÙ [51].
(61) This phenomenon and its implications are specifically discussed in FRASCA - MAROTTA [29].
(62) See COTULA [21], CARANZA - COTTARELLI [10].

articulated financial structure in which risk capital and medium-term securities have re-acquired importance. At the same time, administrative constraints on bank assets (from 1983) and, more gradually, controls on foreign financial operations were lifted (63).

A turn towards a "classical" approach to monetary policy has progressively taken place: money supply has become an intermediate objective (although as target indicator this aggregate is often and mistakenly replaced by total and private sector credit objectives and projections); policy objectives are adopted according to developments of monetary base aggregates; open-market operations are used as a main tool of short-term liquidity management.

In a situation which has seen the development of channels and forms of financing other than those offered by traditional credit institutions, the most significant new feature is the almost complete removal of constraints on international capital movements. Italian households and companies have learned to discriminate investments and sources of financing in different markets and among different currencies on the basis of their yield and costs as well as the range of services offered. On the domestic front, the weight of the public debt, now well over the GDP, on private markets and private sector portfolios has become extremely heavy.

Monetary policy must satisfy the contrasting needs of curbing the cost of the national debt and reducing the inflation rate. Within the EMS fixed exchange rate system, the problem is exacerbated by the growing elasticity of capital movements to interest rates.

Within this context, corporate financial structure has undergone significant changes. The persistence of high positive real interest rates (with respect to foreign rates as well) have shifted a significant amount of resources to the management of circulating capital. In large companies, cash-flow management and financial functions have concentrated and developed. Investments in high-interest securities also begin to provide (for large companies in particular) a significant contribution to companies' operating results as improvements in their balance-sheet accounts lead to the accumulation of considerable financial surpluses. At the same time, direct access to capital markets,

---

(63) See MICOSSI - ROSSI [41].

including foreign markets, helps to significantly reduce the need to resort to credit institutions.

Basically, large companies have now freed themselves from traditional financial institutions and have themselves, to some extent, become financial intermediaries (64). Simultaneous access to diverse markets and financial channels and the availability of considerable financial assets make large companies less exposed to the effects of monetary restrictions. The strength that they have acquired in commercial transactions also puts them in a position of advantage with respect to their commercial partners, both up and down stream.

Medium and small companies, on the other hand, have come out of the restructuring process with margins still high with respect to value-added and assets, but with a weaker production and financial structure. Both the indebtedness with the banking system and interest payments remain high. These are clearly the companies which are finding it more difficult to operate in an environment characterized by "strong" exchange rates and strict monetary policy (65). Since they are more affected by monetary policy changes, they are encountering increasing difficulties in exploiting market opportunities while dealing with technological changes. They often choose to sacrifice efficiency for expansion.

In this situation, the costs of using aggregate monetary measures to control inflation, support the exchange rate and manage the public debt, are bound to rise. The overwhelming problems of medium and small-sized companies are also bound to delay the expansion of the restructuring process which is, instead, extremely urgent in an environment of growing competitive pressure within the single European market.

---

(64) See VACIAGO [54] and other contributions contained in *Economia e politica industriale*, n. 56, October-December 1987.
(65) See DE CECCO [23].

GRAPH 1

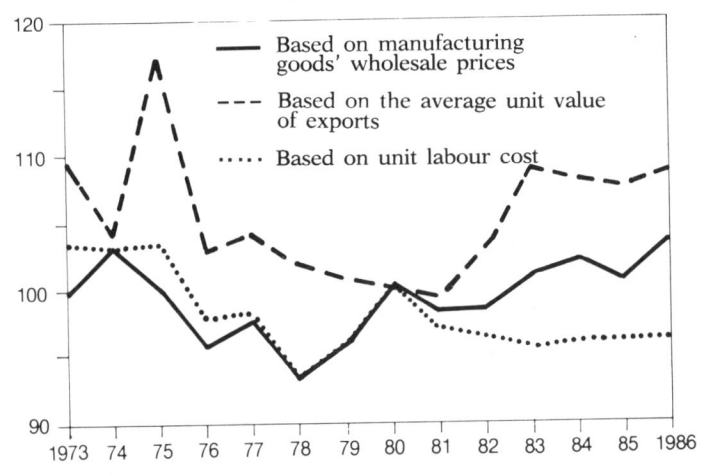

REAL EXCHANGE RATES
(1980 = 100)

—— Based on manufacturing goods' wholesale prices
--- Based on the average unit value of exports
····· Based on unit labour cost

Source: ICE ([35], 1988).

GRAPH 2

INDUSTRY (EXCLUDING CONSTRUCTION): REAL WAGES DEFLATED WITH WHOLESALE (W_1) AND CONSUMER (W_2) PRICE INDICES
(1970 = 100)

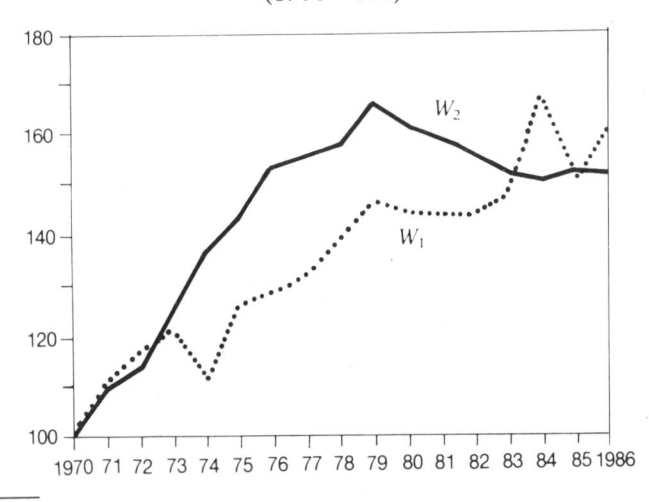

Source: ISTAT; our calculations.

GRAPH 3

**INDUSTRY (EXCLUDING CONSTRUCTION):
EFFECTIVE GROSS PROFIT MARGINS (P_1)
AND SIMULATED MARGINS AT CONSTANT SOCIAL SECURITY
CONTRIBUTIONS/WAGES RATIO (P_2)**

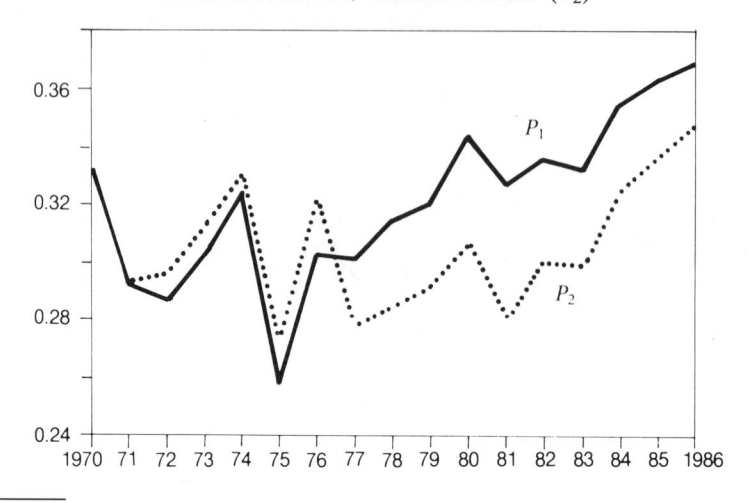

Source: ISTAT; our calculations.

GRAPH 4

INTEREST RATES AND WHOLESALE PRICES
(in percent)

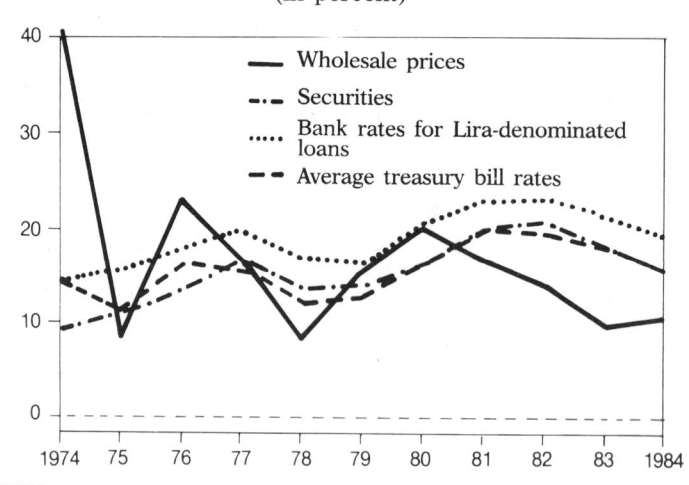

— Wholesale prices
-·- Securities
····· Bank rates for Lira-denominated loans
-- Average treasury bill rates

Source: BANCA D'ITALIA [3] and ISTAT.

The Role of Monetary and Financial Policies etc. 103

GRAPH 5

INDUSTRIAL COMPANIES' INDEBTEDNESS
(in percent)

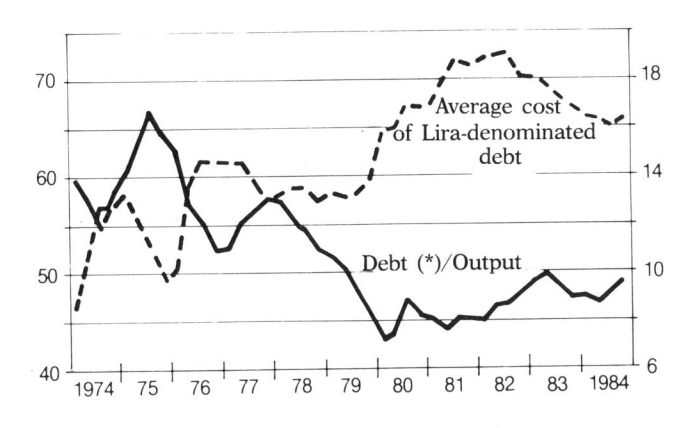

(*) average seasonally-adjusted stock of debt
Source: BANCA d'ITALIA [3].

GRAPH 6

INDUSTRIAL COMPANIES' FINANCIAL BURDENS
(in percent)

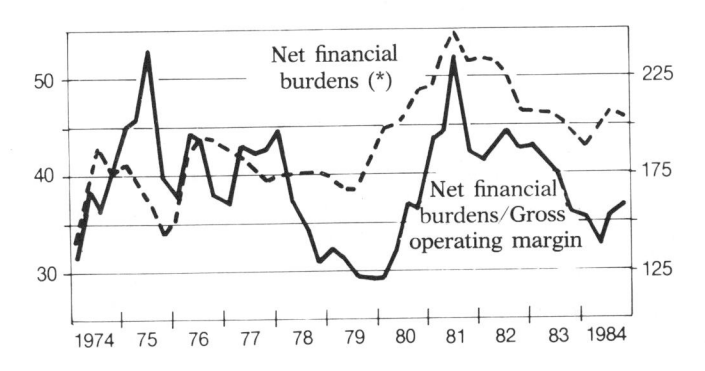

(*) Constant price index (1970 = 100).
Source: BANCA d'ITALIA [3].

GRAPH 7

**MANUFACTURING INDUSTRY:
GROSS PROFIT MARGINS/VALUE-ADDED (IN PERCENT)
PER CLASS OF EMPLOYEES**
(1975 = 100)

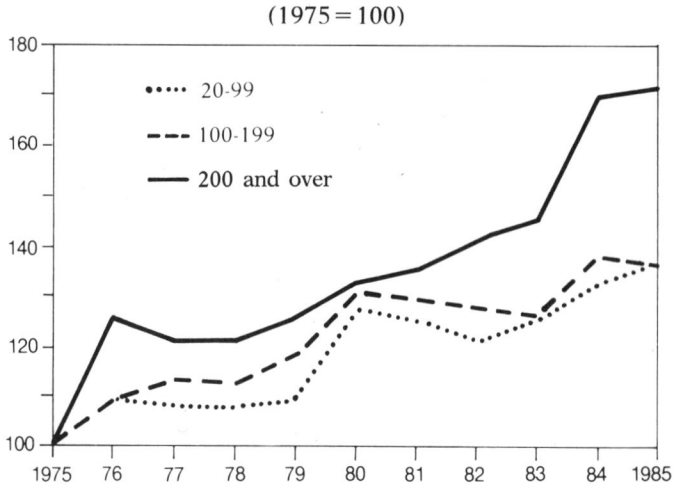

••••• 20-99
--- 100-199
—— 200 and over

Source: ISTAT: *Indagine sul prodotto lordo*; our calculations.

GRAPH 8

**MANUFACTURING INDUSTRY:
VALUE-ADDED/WORKERS**
(1975 = 100)

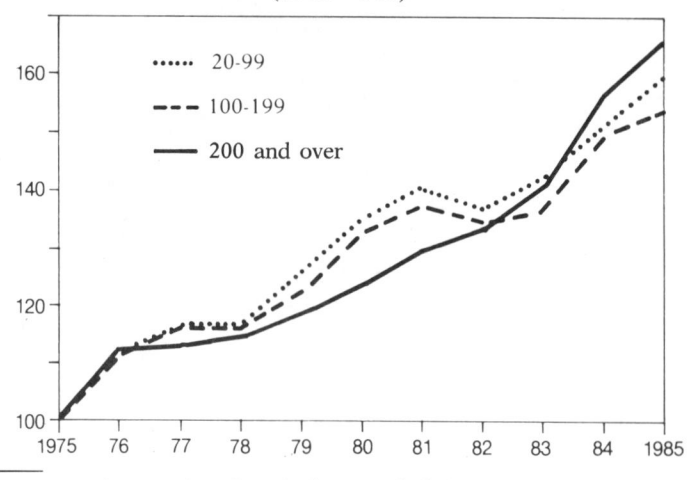

••••• 20-99
--- 100-199
—— 200 and over

Source: ISTAT: *Indagine sul prodotto lordo*; our calculations.

The Role of Monetary and Financial Policies etc. 105

GRAPH 9

**MANUFACTURING INDUSTRY:
FIXED INVESTMENT AT 1980 PRICES PER CLASS OF EMPLOYEES**
(1980 = 100)

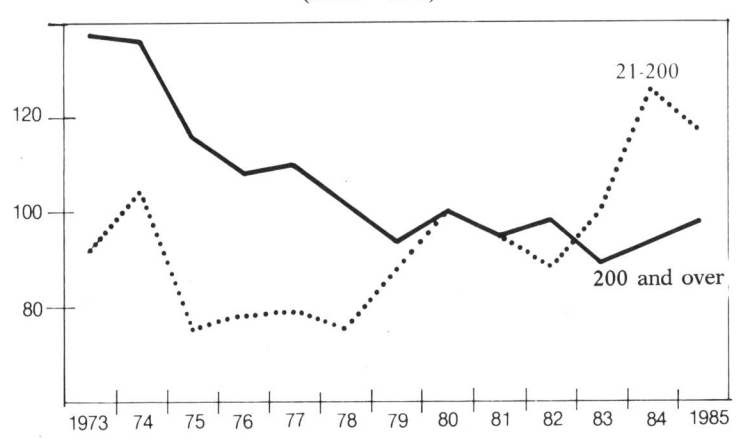

Source: BARCA - MAGNANI [5].

GRAPH 10

**MANUFACTURING INDUSTRY:
FINANCIAL LIABILITIES/TOTAL ASSETS
PER CLASS OF EMPLOYEES**
(1973 = 100)

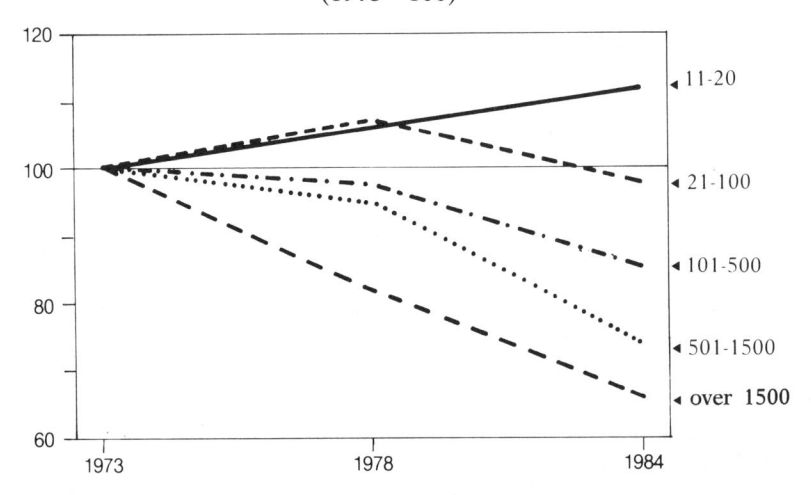

Source: MEDIOCREDITO CENTRALE [39]; our calculations.

GRAPH 11

**GROSS ASSETS (A)
AND SHORT-TERM FINANCIAL DEBT (D_1)
PER SIZE CLASSES (*)**
(indices 1982 – 1986)

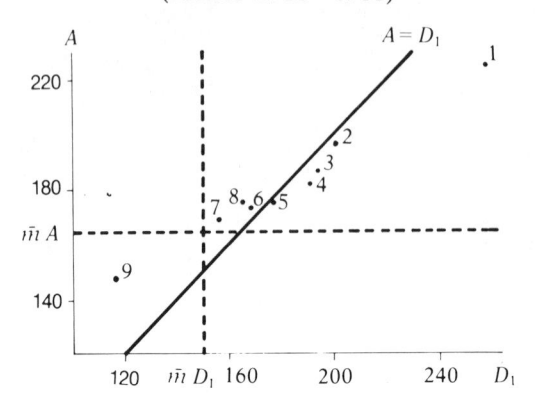

(*) See note to Chart 12.
Source: CENTRALE DEI BILANCI; our calculations.

GRAPH 12

**GROSS ASSETS (A)
AND MEDIUM LONG-TERM DEBT (D_2)
PER SIZE CLASSES (*)**
(indices 1982 – 1986)

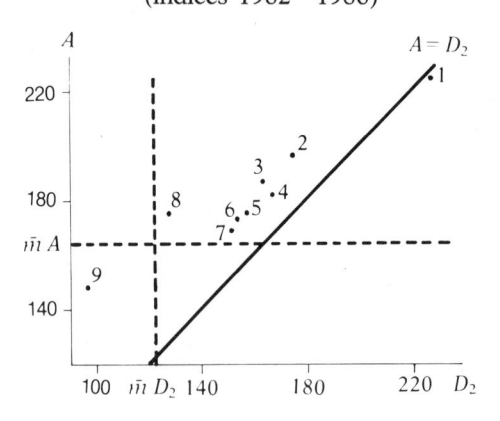

(*) Classes of employees: 1 = 1-10   4 = 41-80   7 = 321-640
                           2 = 11-20   5 = 81-160   8 = 641-1280
                           3 = 21-40   6 = 161-320   9 = above 1280

Source: CENTRALE DEI BILANCI [12]; our calculations.

BIBLIOGRAPHY

[1] BANCA COMMERCIALE ITALIANA: «La strategia finanziaria delle imprese industriali negli anni '80: le nuove determinanti della domanda di impieghi bancari», *Tendenze reali*, n. 28, July 1985.
[2] ——: «Esiste nel breve periodo un vincolo di potenziale produttivo?», *Tendenze reali*, n. 33, July 1987.
[3] BANCA d'ITALIA: *Relazione annuale*, various years.
[4] BARBONE L. - BODO G. - VISCO I.: «Costi e profitti nell'industria in senso stretto: un'analisi su serie trimestrali», *Bollettino della Banca d'Italia*, 1981.
[5] BARCA F. - MAGNANI M.: *L'industria fra capitale e lavoro*, Bologna, il Mulino, 1989.
[6] BIANCHI B.: «I tassi d'interesse sui mercati monetario e finanziario», in COTULA F. - de' STEFANI P. (eds.): *La politica monetaria in Italia*, Bologna, il Mulino, 1979.
[7] BIANCHI P.: «Produzione meccanica e automazione industriale», *L'Industria*, n. 4, October-December 1983.
[8] BISCAINI A.M. - COTULA F. - NARDI P. - ZECCHINI S.: «Le istituzioni creditizie in Italia», in COTULA F. - de' STEFANI P. (eds.): *La politica monetaria in Italia*, Bologna, il Mulino, 1979.
[9] CANOVI L.: «La distribuzione del credito ordinario e speciale ai settori dell'industria manifatturiera (1973-1984)», in ONADO M.: *Sistema finanziario e industria*, Bologna, il Mulino, 1986.
[10] CARANZA C. - COTTARELLI C.: «Financial Innovation in Italy: a Lopsided Process», Roma, Banca d'Italia, *Temi di discussione*, n. 64, May 1986.
[11] CARANZA C. - FAZIO A.: «Methods of Monetary Control in Italy: 1974-83», in HODGMAN D. (ed.): *The Political Economy of Monetary Policy: National and International Aspects*», Boston, Fed. Reserve Bank, 1983.
[12] CENTRALE DEI BILANCI: *Economia e finanza dell'industria italiana*, Torino, Centrale dei Bilanci, 1988.
[13] CER: «Decentramento e competitività dell'industria italiana», *Rapporto*, n. 6, Roma, Cer, 1984.
[14] CIAMPI C.A.: «Intervento al XII congresso nazionale del Forex Club italiano, 28 ottobre», Roma, Banca d'Italia, *Bollettino economico*, November 1989.
[15] CIOCCA P. - FRASCA F.: «I rapporti fra industria e finanza: problemi e prospettive», *Politica economica*, n. 1, April 1987.
[16] CIPOLLETTA I.: «Le politiche congiunturali e la ristrutturazione produttiva», in CIPOLLETTA I. (ed.): *Struttura industriale e politiche macroeconomiche in Italia*, Bologna, il Mulino, 1986.
[17] CIPOLLETTA I. - HEIMLER A. - CALCAGNINI G.: «Ristrutturazione e adattamento dell'industria italiana», *Economia italiana*, n. 3, September-December 1987.
[18] CIPOLLETTA I. - HEIMLER A.: «Processi di ristrutturazione, progresso tecnologico e crescita economica», *Rivista di Politica economica*, n. 7-8, July-August 1989.
[19] CONTI V.: «Replica e alcune osservazioni sul processo di ristrutturazione delle imprese negli anni '80», in Atti del seminario: Ristrutturazione economica e finanziaria delle imprese, Roma, Banca d'Italia, *Contributi all'analisi economica*, special issue, 1988.
[20] CONTINI B.: «Grandi e piccole imprese industriali in Italia: dinamica e performance negli anni '80 a confronto», *Moneta e credito*, n. 163, July-September 1988.

[21] Cotula F.: «Innovazione finanziaria e controllo monetario», *Moneta e credito*, n. 145, January-March 1984.

[22] Cotula F. - Morcaldo G.: «Attività e passività finanziarie degli utilizzatori finali delle risorse in Italia», in Cotula F. - de Stefani P. (eds.): «*La politica monetaria in Italia*», Bologna, il Mulino, 1979.

[23] De Cecco M.: «Commento al X rapporto del Centro studi Confindustria», *Rivista di politica economica*, n. 8-9, August-September 1988.

[24] Delli Gatti D. - Saraceno P.: «Accumulazione di capitale e ristrutturazione industriale: una riconsiderazione del «caso italiano» negli ultimi venti anni», in Barbetta G.P. - Silva F. (eds.): *Trasformazioni strutturali delle imprese italiane*, Bologna, il Mulino, 1989.

[25] Fanna A. - Papadia F. - Salvemini G.: «Attività e passività finanziarie degli utilizzatori finali delle risorse in Italia», in Cotula F. (ed.): *La politica monetaria in Italia*, Bologna, il Mulino, 1989.

[26] Fazio A.: «La politica monetaria in Italia dal 1947 al 1978», *Moneta e credito*, n. 127, September 1979.

[27] Filippi E.: «Imprese e inflazione: considerazioni sulla politica industriale italiana dell'ultimo decennio», *L'industria*, n. 3, July-September 1983.

[28] Frasca F. - Marotta G.: *La ristrutturazione finanziaria della medio-grande impresa*, unpublished manuscript, Roma, Banca d'Italia, 1987.

[29] ——·——: «La ristrutturazione finanziaria delle grandi imprese», in Atti del seminario «Ristrutturazione economica e finanziaria delle imprese», Roma, Banca d'Italia, *Contributi all'analisi economica*, special issue 1988.

[30] Giavazzi F. - Giovannini A.: «The Role of the Exchange Rate Regime in a Disinflation: Empirical Evidence on the EMS», in Giavazzi F. - Micossi S. - Miller M. (eds.): *The European Monetary System*, Cambridge, Cambridge University Press, 1988.

[31] Giavazzi F. - Spaventa L.: «Italy: the real effects of inflation and disinflation», *Economic Policy*, n. 8, April 1989.

[32] Gressani D. - Guiso L. - Visco I.: «Il rientro dall'inflazione: un'analisi con il modello econometrico della Banca d'Italia», *Temi di discussione*, n. 90, July 1987.

[33] Gros-pietro G.M. - Rosa G.: «Investimenti, processi innovativi e riflessi sulle strategie di impresa», *Rivista di politica economica*, n. 2, February 1987.

[34] Heimler A. - Milana C.: *Prezzi relativi, ristrutturazione, produttività*, Bologna, il Mulino, 1984.

[35] Ice: *Rapporto sul commercio estero*, Roma, Ice, 1987 and 1988.

[36] Laboratorio di Politica Industriale: *Rapporto sulle tendenze della ristrutturazione industriale in Italia*, Bologna, Nomisma, 1983.

[37] Landi A.: «Redditività e grado di indebitamento delle imprese industriali», in Onado M. (ed.): *Sistema finanziario e industria*, Bologna, il Mulino, 1986.

[38] Macchiati A.: *Il finanziamento delle imprese industriali in Italia*, Bologna, il Mulino, 1985.

[39] Mediocredito Centrale: *Indagine sulle imprese manifatturiere*, Roma, Mediocredito Centrale, 1973, 1978, 1984.

[40] Micossi S. - Rebecchini S.: «L'efficacia degli interventi sul mercato dei cambi: il caso della lira italiana», *Rivista di politica economica*, n. 5, May 1984.

[41] Micossi S. - Rossi S.: «Controlli sui movimenti di capitale in Italia», *Giornale degli economisti*, n. 1, January-February 1986.

[42] MONTI M. - CESARINI F. - SCOGNAMIGLIO C.: *Il sistema creditizio e finanziario italiano*, Roma, Ministero del Tesoro, 1982.
[43] ONADO M.: «Il finanziamento delle imprese. Un tentativo di sintesi», in ONADO M. (ed.): *Sistema finanziario e industria*, Bologna, il Mulino, 1986.
[44] ONOFRI P. - SALITURO B.: «Inflazione e politiche di stabilizzazione in Italia», *Politica economica*, n. 2, August 1985.
[45] PONTOLILLO V.: «L'intervento dello Stato sul costo del credito» in COTULA F. - de' STEFANI P. (eds.): *La politica monetaria in Italia*, Bologna, il Mulino, 1979.
[46] RIVA A. - SILVA F.: *Scelte strategiche e riorganizzazione aziendale: la grande impresa nel decennio 1975-1984*, Milano, F. Angeli, 1985.
[47] SEMBENELLI A.: «Effetti dell'inflazione sulla situazione finanziaria delle imprese: il caso italiano», in ZANETTI G. (ed.): *Alle radici della struttura produttiva italiana*, Roma, Sipi, 1988.
[48] SIRACUSANO F. - TRESOLDI C.: «Evoluzione e livelli dei margini di profitto dell'industria in Italia e nei principali paesi industriali», in Atti del seminario Ristrutturazione economica e finanziaria delle imprese, Roma, Banca d'Italia, *Contributi all'analisi economica*, special issue 1988.
[49] SPAVENTA L.: «Two Letters of Intent; External Crises and Stabilization Policy, Italy 1973-77», in WILLIAMSON J. (ed.): *Imf conditionality*, Washington, Institute for International Economics, 1983.
[50] TARANTOLA RONCHI A.M.: «Il mercato monetario», in COTULA F. - de' STEFANI P. (eds.): *La politica monetaria in Italia*, Bologna, il Mulino, 1979.
[51] TRAÙ F.: «Risorse finanziarie, composizione degli investimenti e dimensione delle imprese nell'industria manifatturiera: 1968-84», *Rassegna economica*, n. 6, November-December 1987.
[52] TRAÙ F.: «Dimensione di impresa e struttura dei tassi di profitto nell'industria manifatturiera italiana: un'analisi disaggregata», *L'industria*, n. 4, October-December 1988.
[53] TRAÙ F.: «L'evoluzione dimensionale delle imprese manifatturiere italiane negli anni 1982-1986», *Rassegna economica*, n. 1, January-March 1990.
[54] VACIAGO G.: «Novità e problemi nei rapporti tra industria e finanza», *Economia e politica industriale*, n. 56, October-December 1987.
[55] ZANETTI G.: «Mutamenti recenti e prospettici nella natura dell'accumulazione», *Note economiche*, n. 5-6, November-December 1985.
[56] ——: «Fasi di sviluppo, profili decisionali e cambiamento tecnico: l'esperienza italiana», Conference organized by Ceris on *Analisi dello sviluppo d'impresa*, Torino, 26 November 1987.

Disinflation in Italy:
An Analysis with the Econometric Model of the Bank of Italy (*)

Daniela Gressani (**) - **Luigi Guiso - Ignazio Visco**
Banca d'Italia, Roma-World Bank Washington   Banca d'Italia, Roma

1. - Introduction

At the end of 1986, the annual inflation rate in Italy, measured in terms of the consumer price index, was 4.2%; wholesale prices actually fell by 2.5%. The inflation rate had not been so low since before the first oil shock.

A significant part of this favorable outcome can be attributed to the large drop in oil prices in early 1986. Even prior to 1986, however, the inflation rate in terms of consumer prices had steadily declined from the peak of 22%, reached in November 1980 after the second oil shock, to 8.8% by December 1985.

This earlier reduction was the result of a complex process aimed at bringing inflation under control, which involved some notable behaviors by private parties as well as government policies: the entry into the European Monetary System; a steady and restrictive monetary policy; the self-restraint of wage earners, within the broader

---

(*) Reprinted by permission of Elsevier Science Publishing Co., Inc. from article, by author: *Journal of Policy Modeling*, vol. 10, n. 2, pp. 163-203. Copyright 1988 by the Society for Policy Modeling. The authors are grateful to Piero Rubino for productive discussions and to Albert Ando and Giampaolo Galli for useful suggestions. The views presented in this paper are those of the authors and do not necessarily reflect those of the Banca d'Italia.
This paper was published in Italian, Roma, Sipi, Series *Monografie RPE*, vol. no. 8, 1991.
(**) Since 1989 the World Bank, Washington (DC).
Advise: the numbers in square brackets refer to the Bibliography in the appendix.

framework of a policy of "precommitment" to target inflation rates announced by the government, and the management of publicly regulated prices tied to this target.

In this paper the quarterly econometric model of the Bank of Italy (Banca d'Italia [1]) is used to provide a systematic description of this disinflation process and to identify factors which obstructed it. First, in Section 2 the most relevant events and policy measures of the period are briefly reviewed. In Section 3 the need to endogenize, however crudely, some policy instruments as well as some international variables is considered. After examining the basic structure of the price-wage sector of the model, the simulation exercises conducted to evaluate the nature of the disinflation process are then illustrated.

In Section 4, on the basis of the simulation, what part of Italian inflation (and disinflation) can be attributed to the "rest of the world" is estimated, accounting for the oil shock of 1979, the dollar appreciation of 1981-1985, the negative oil shock of early 1986, and the almost simultaneous depreciation of the dollar. The contribution of monetary and exchange rate policies to the disinflation process is then examined. Finally, an attempt to evaluate the effect of the policy of "precommitment" (1983-1984) for wages and publicly regulated prices is presented.

2. - Disinflation 1980-1986: Facts and Policies

In this section the main economic developments that accompanied the evolution of domestic prices after the second oil shock are briefly reviewed. This survey will be confined to those variables most directly related to inflation (1).

2.1 *The Variables at Play*

Examining the behavior of the consumer price index in the 1970s reveals three moments of sudden acceleration in inflation, corres-

---

(1) For further information on the facts and policies that took place over the years considered in this paper, See BANCA D'ITALIA *Annual Report*, 1986.

ponding to the two oil crises and the exchange rate crisis of 1976-1977 (Graph 1).

After reaching the peak of 21.8% in the third quarter of 1980 as the second oil shock took place, the annual rate of inflation began a slow deceleration. This phase was marked by several changes of direction: four years were necessary to halve the rate of inflation, from 21.2% in 1980 to 10.8% in 1984. In 1985 the disinflationary process halted: the reduction achieved in terms of the annual rate of inflation (down to 9.2%) was determined solely by the statistical effect of the decline that occurred at the end of 1984. Finally in 1986, with the negative oil shock and the steady depreciation of the dollar, inflation fell below 6%.

The growth rate of manufactured goods prices declined to a larger extent, on average, and more steadily than the growth rate of consumer prices. In fact, the latter reflected the diverse developments

GRAPH 1

WHOLESALE AND CONSUMER PRICE INDICES
(annual percentage changes on yearearlier period)

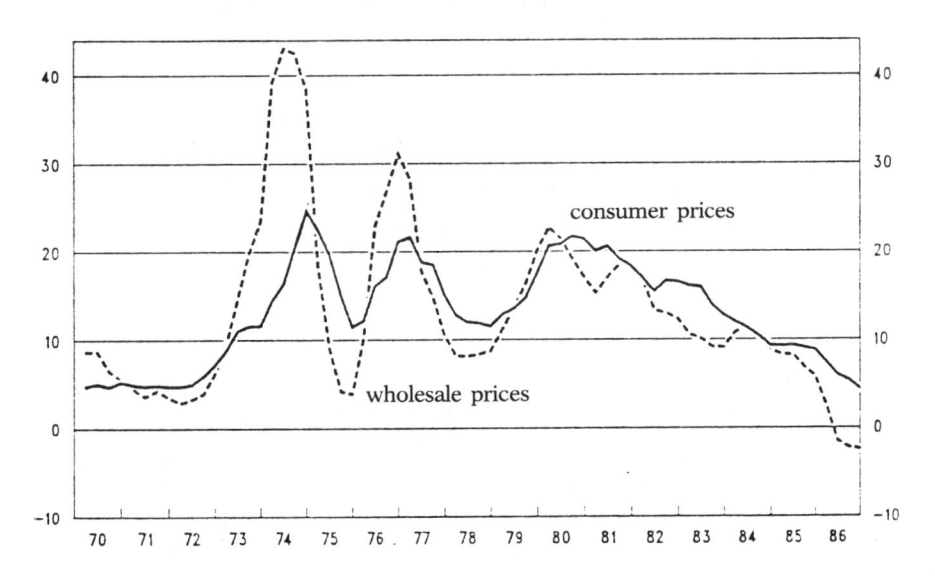

Source: ISTAT: *Bollettino mensile di statitistica,* various issues.

of publicly regulated prices and the improvement in the terms of trade between the service and the manufacturing sectors, which became more pronounced during the 1980s. The need to reconcile balancing the budget of public utilities with the announced targets for inflation induced the policy makers to differentiate their actions on regulated prices according to cyclical conditions and the price aggregate taken as reference. Between 1980 and 1984, while self-restraint initiatives were promoted in the manufacturing and retail sectors, publicly regulated prices (inclusive of rents and other prices subject to government supervision) grew at higher rates than consumer prices. In the most recent years, on the contrary, stronger inflationary impulses originated from goods and particularly from services not subject to government control.

Large excess supply of raw materials in international markets, owing to the lasting recession in developed countries, caused dramatic drops in the prices of raw materials after 1981. Regarding oil, the presence of a producers' cartel succeeded, however effortfully, at preventing large falls in oil prices until 1985, although the price of oil imported by OECD countries decreased from $36.5 to $27.5 between 1980 and 1985. In the summer of 1986, as Saudi Arabia abandoned its role of swing producer and sizable quantities of oil reached the market from areas outside the cartel, oil prices fell below $10 per barrel, recovering slightly at the end of the year. The average cif price of oil imported by Italy that year was approximately $14 per barrel. In spite of the drop of international prices, the lira prices of imported raw materials and semifinished inputs underwent a remarkable increase between 1980 and 1985, due to the steady appreciation of the dollar.

After a strong increase in 1981, real before-tax wage rates per employee grew less than 1 percent in 1982-1983 and in 1985. Benefiting from the improvement in industrial relations, the cyclical upturn in 1983, and the rise in capacity utilization, profit margins in the manufacturing sector started to rise in 1984, after the contraction that had been taking place since 1981. This rise in profit margins contributed to the high growth rates of the output price of manufactured goods. However, the evolution of the output price of manufactured goods did not prevent the progressive reduction of inflation differentials with competing countries. This reduction was not sufficient to keep the level of relative prices unchanged; in spite of the depreciation of the nominal exchange rate of the lira, the manufacturing sector suffered a loss of competitiveness of more than 3 percentage points between 1980 and 1985 (Graph 2).

2.2 *The Policies*

These developments occurred while important changes in the economic policies were taking place. In particular, the *exchange rate policy* aimed mainly at maintaining the "fixed" parity imposed by the European Agreement of 1979. The other instruments on which the policy makers relied most heavily in the attempt to attain an acceptably rapid disinflation consisted of *monetary policy, incomes policy* — in the form of *precommitment* to a given growth of the cost-of-living index underlying the working of the indexation system (the so-called "scala mobile") — and *publicly regulated prices policy*.

In most recent years, monetary policy has been gradually overcoming the uses of administrative measures as instruments for the control of credit in favor of a system of indirect controls on monetary aggregates. Between the end of 1980 and the first quarter of 1981, in the presence of exchange rate tensions and accelerating inflation, the discount rate was gradually raised from 15 to 19%. The higher degree of restrictiveness in monetary policy resulted also in the tightening of credit ceilings and the extending of the obligation to invest a given share of commercial banks' assets in Treasury bills — thus causing a rapid increase in real interest rates (Graph 3).

In July 1981, the Bank of Italy abandoned the role of residual buyer of Treasury bills issued to finance the budget deficit — the so-called "divorce". The Bank thus increased its degree of independence in managing the monetary base and its ability to control the composition of the public sector deficit financing. Since April 1983, the end of the most severe phase of inflation, nominal interest rates began to decline as inflation was slowing down, consolidating the policy rule aimed at maintaining the real rate constant.

In the 1980s, incomes policy and social pact policy interwove with the developments of industrial relations. In December 1981,

GRAPH 2

ITALY'S EFFECTIVE REAL AND NOMINAL EXCHANGE RATES AND RELATIVE PRICES (*)

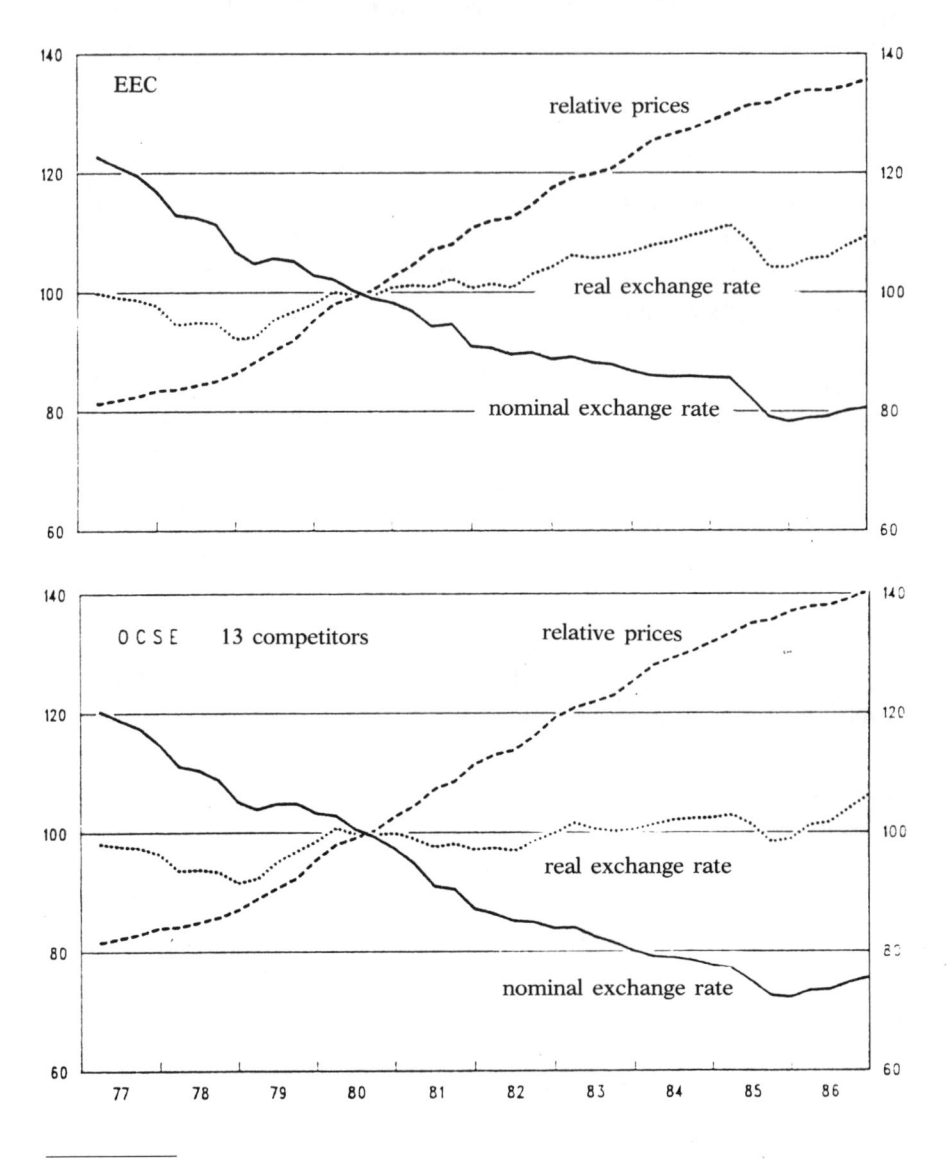

(*) Indices computed on the basis of wholesale prices of manufactured goods (including oil products) relative to the average for the EEC and the 13 main competitors, 1980 = 100.

Source: BANCA D'ITALIA: *Economic Bulletin*, various issues.

GRAPH 3

INTEREST RATES ON TREASURY BILLS

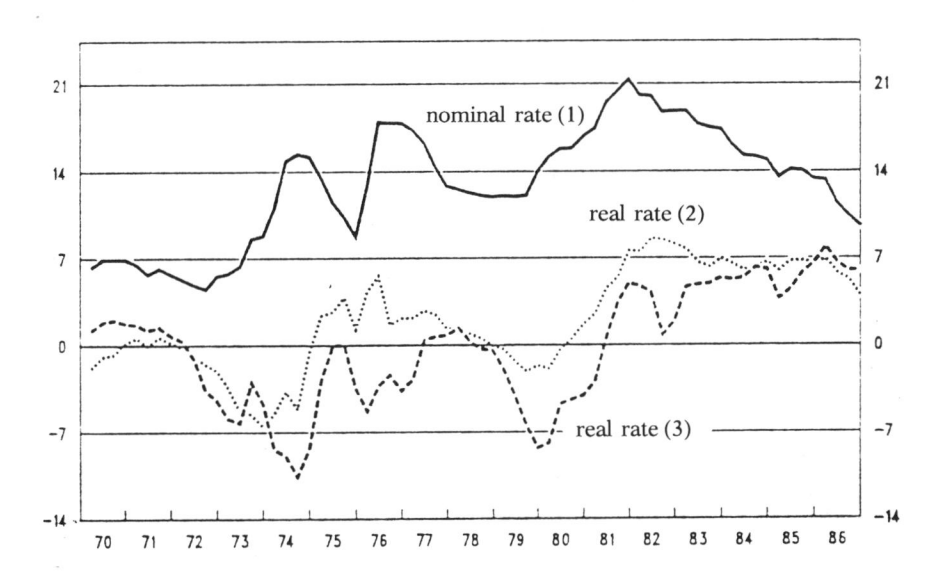

(1) Weighted average of 3, 6, and 12 month Treasury Bills;
(2) Deflated with expected consumer price changes (*Mondo economico* surveys);
(3) Deflated with actual consumer price changes.
Source: BANCA D'ITALIA: *Economic Bulletin*, various issues.

trade unions put forward a joint proposal on labor costs. In exchange for a commitment to restraining wage growth below 16% in 1982 (when the closing of new labor contracts was due), the government was asked to adopt a set of fiscal and social security measures; furthermore, it was proposed to extend the target growth of 16% to publicly regulated prices.

The following year, in an embittered social climate owing to the difficult collective bargaining, Confindustria (the manufacturing sector trade association) unilaterally renounced the 1975 agreement on the *scala mobile* system. An overall agreement was reached later, following an initiative by the government, covering taxation, wages and regulated prices (*Protocollo d'intesa* of January 1983). The agreement included the announcement by the government of its commitment to limit the growth rate of consumer prices to a target of 13% in 1983 and 10% in 1984. In exchange for the commitment by the government to present a bill lowering fiscal pressure to support employment opportunities for the young, the trade unions and the trade associations agreed to fix ceilings for wage increases when closing the labor contracts for the period 1983-1985. Furthermore, to reduce the transmission of inflation through wage indexation, the computation of the indexation provisions was modified, causing a 15% reduction in the degree of protection of wages and salaries from inflation.

In early 1984, the inability of the trade unions to reach a common agreement regarding the reform of wage indexation mechanisms induced the government to independently propose a bill, according to which the increase in the *scala mobile* cost-of-living index was set in advance to be equal to 9 points during 1984 (2). The following parliamentary passage of the bill limited its effectiveness to the first semester of 1984; nevertheless, the weakening of inflation made it possible to attain the target inflation rate of 10%. While the restoration of the indexation provisions following the increase in the cost-of-living index, accrued but not paid in 1984, was rejected by referendum, the trade unions and associations again failed to reach an agreement on reforming the indexation system in the private sector. The possibility of extending to all employees the indexation mechanism (introduced by the agreement of December 1985) concerning the public sector was then prospected, and the new system was enacted by the government in January 1986 (3).

The presence of a significant share of publicly regulated prices in the aggregate consumer price index allows the economic authorities

---

(2) According to the *scala mobile* system then in effect, wages were adjusted to changes of the cost-of-living index every 3 months. The adjustment was equal to the absolute change of the index times a fixed amount, independent of actual wages. See Visco [16] for further details.

(3) The most relevant characteristics of the new system consist of decreasing the frequency of the cost-of-living adjustments to twice per year and in differentiating the degree of indexation for two types of earnings: minimum wages which are fully indexed and wages and salaries above the minimum, whose degree of indexation is less than proportional. For further details, See BANCA D'ITALIA: *Economic Bulletin*, n. 2, 1986, pp. 55-61.

to directly affect its developments (4). Policy makers have continually resorted to regulated prices policy in the most recent years, aiming at a difficult equilibrium between improving the budget of public utilities, much deteriorated during the 1970s, and contributing to bringing the inflation rate down. The goal of intervention was identified as limiting the rate of growth of publicly regulated prices within the target inflation rate. Regulated and "supervised" prices, according to the agreement of 1983, and regulated prices and rents (5), according to the agreement of 1984, were the aggregates monitored by the government in recent years.

3. - Conditional Simulation Analysis with the Econometric Model

The main goal of this paper is to quantitatively evaluate the determinants of the disinflation process that occurred in Italy during the early 1980s. The Bank of Italy quartely econometric model is used for this purpose. Before presenting the simulation exercises, it is necessary to discuss some general issues.

3.1 *The Reaction Functions*

Even assuming that the estimated structure is relatively invariant with respect to changes in policies and other exogenous variables — at least to changes whose order of magnitude is similar to those actually experienced (on this issue Sims' comments on Lucas [11], in Sims [13], pp. 118-9) — nevertheless, a problem concerning the treatment of exogenous variables arises in conditional simulations with an

---

(4) The share of goods and services subject to price supervision by the government, computed in terms of 1970 lire, can be estimated around 30% of total consumption by the household sector according to the input-output table for 1978. The total input-output elasticity of the consumer price index with respect only to government regulated prices of public services and electricity has been estimated about 15%, one third of which being due to their nature of inputs to the production process (see RUBINO-VISCO [12]).

(5) At the end of 1983, the annual percentage change of rents had already exceeded the target for 1984. The government then suspended the inflation-linked rise in rents for 1985, mandated by the law on "fair rent" (*equo canone*).

econometric model. That is, even if for some variables, e.g., policy instruments, one can assume weak exogenity (ENGLE, HENDRY, and RICHARD [4]), a concept pertaining to the domain of parameter estimation, these variables must be strongly exogenous in order to assume in conditional simulation that they can be maintained at their historically observed values.

In order to lessen, to some extent, this problem, a number of policy instruments have been endogenized. For both the exchange rate policy and the monetary policy, reaction functions have been used to allow for policy responses to changes in endogenous variables. For the former, the function is the one estimated in the current version of the econometric model (See equation 9.64.A, vol. II, discussed in vol. I, pp. 286-9). This function determines the effective exchange rate (EXCH) and, for given exogenous cross-rates between the currencies other than the lira, the bilateral rates are obtained. Table 1, in addition to the estimates for the period 1973.3-1983.4 used in the exercises, presents estimates for sample periods covering more recent years, including 1986. They all indicate a remarkable stability of the parameters.

The equation is estimated in logs; the most relevant explanatory variables are the price differential between Italy and her main trade partners and the ratio of short-term foreign currency reserves to a moving average of imports. Although the estimation period starts in 1973, the transition point to the flexible exchange rate regime, the regressions actually reflect behaviors followed after the enactment of the European Monetary Agreement (the observations for the period 1975-1976, when a major crisis in the exchange market occurred, have in fact been corrected with a dummy). Moreover, since the second half of 1979 and additional explanatory variable, the change in the dollar/deutsche mark rate, is introduced to capture the short-term effects that, according to the monetary agreement, are produced on the lira exchange rate when there are fluctuations in the dollar/deutsche mark.

The relative price variable has a low impact elasticity, approximately 10%, and a long run elasticity that is slightly smaller than unity: this reflects the sluggishness with which the nominal exchange rate has accommodated losses of competitiveness, relative to Italy's

TABLE 1

REACTION FUNCTION OF THE EFFECTIVE EXCHANGE RATE (*)

T	1983.4	1984.4	1985.4	1986.4
c_0	−0.143	−0.103	−0.124	−0.110
	(1.769)	(1.461)	(1.727)	(1.593)
c_1	0.109	0.080	0.096	0.085
	(1.882)	(1.587)	(1.864)	(1.730)
c_2	0.100	0.098	0.075	0.117
	(1.755)	(1.852)	(1.611)	(2.990)
c_3	0.880	0.905	0.893	0.898
	(14.984)	(17.197)	(16.518)	(17.056)
c_4	−0.014	−0.014	−0.014	−0.013
	(3.192)	(3.385)	(3.369)	(3.139)
c_5	0.053	0.054	0.054	0.054
	(11.743)	(12.383)	(11.863)	(11.895)
R^2	0.998	0.998	−0.988	−0.998
SER	0.011	0.011	0.011	0.011
DW	2.041	2.023	1.942	1.903
MLM1	0.030	0.022	0.020	0.099
MLM4	0.156	0.137	0.004	0.008
ARCH1	1.279	1.088	1.375	1.356
N	42	42	42	42
CHOW	1.074	0.737	1.082	1.127

(*) Estimated equation:
$$\log(EXCH) = c_0 + c_1 \log(PQMFD/PREST) + c_2 \log(EXDMUS_{-1})(1 - DUB793)$$
$$+ c_3 \log(EXCH_{-1}) + c_4 \log(RUNOR / \sum_{i=0}^{3} IMP_{-1})$$
$$+ c_5 (DU761 + DU762 + DU763 - DU752 - DU753 - DU754 - DU733).$$

Estimation period: $1973.3 - T$.

Variables: $DUBXYZ$ = dummy (=1 prior to 19XY.Z): $DUXYZ$ = dummy (=1 in 19XY.Z); $EXCH$ = lira effective exchange rate: $EXDMUS$ = deutsche mark/dollar exchange rate; IMP = imports of goods and services (seasonally unadjusted); $POMFD$ = output price, manufacturing sector; $PREST$ = index of the foreign price of manufactured goods (weighted with trade shares); and $RUNOR$ = short term foreign currency reserves, excluding gold.

Legend:
R^2 = determination coefficient;
SER = standard error of the regression;
DW = Durbin-Watson statistic;
MLM1 = modified Lagrange multiplier test for residuals following simple AR or MA processes of order $i = 1,4$; asymptotically distributed as F with 1, $N-K-1$ degrees of freedom, where N is the number of observations and K the number of regressors;
ARCH1 = conditional autoregressive heteroskedasticity test; computed for a first-order autoregressive process in the residuals and asymptotically distributed as chi square with 1 degree of freedom;
N = number of observations;
CHOW = test of predictive stability, performed with respect to estimates over a sample period ending in 1982.4; distributed as F with R, $N-K$ degrees of freedom, where N is the number of observations, K the number of regressors and R the number of quarters between 1983.1 and T.

Ordinary least squares estimates.
The absolute values of the t-statistics are reported in parentheses under the estimated coefficients.

Source: BANCA D'ITALIA, DATA BANK: *Quarterly Econometric Model* (as of April 1987).

trade partners, generated by different rates of inflation (Graph 2). The negative effect of the ratio of foreign currency reserves to imports reflects exchange rate adjustments that have followed persistent imbalances.

In essence, the exchange rate rule is parametrically described by:

$$(1) \qquad e = \alpha_0 + (1-\alpha_1)\alpha_2(p-p^*) + \alpha_1 e_{-1} - (1-\alpha_1)\alpha_3 x,$$

where e is the logarithm of the effective exchange rate of the lira, p and p^* are, respectively, the logarithms of the price of domestic manufactured goods and the price of manufactured goods produced by Italy's competitors, and x is ratio of foreign currency reserves to imports. With $\alpha_1 = 0$, $\alpha_2 = 1$, and $\alpha_3 = 0$, a policy is defined that fully accommodates the price differentials relative to the rest of the world. With $\alpha_1 = 1$ and $\alpha_0 = 0$, a regime is obtained where the nominal exchange rate is kept at an arbitrarily fixed level $e_{-1} = \bar{e}$. The closer α_2 is to 0, the more restrictive the exchange rate policy.

As far as monetary policy is concerned, a reaction function is estimated for the short-term Treasury bills rate. After the rapid increase that followed the exchange rate crisis of early 1976, the average interest rate on short-term Treasury bills declined gradually from a peak of 18% to approximately 12%. It was then maintained at this value, experiencing only small fluctuations, from the fourth quarter of 1977 to the third quarter of 1979 (Graph 3). In the meantime, the increase of the inflation rate following the second oil shock had gradually determined a large reduction in real rates, which were negative during all of 1979. Starting from the fourth quarter of that year, the rate on Treasury bills (and the discount rate) was gradually but steadily raised, until it reached a peak above 21% at the end of 1981. Measured in terms of the expected inflation rate of consumer prices, the corresponding real rate went from -2.1% in 1979.3 to 2.5% in 1981.1 (a level equal to the average rate existing between 1975.1 and 1978.2), to a maximum of 8.6% in 1982.2. Since then and until 1986, with the exception of a slight decline between the end of 1982 and early 1983, the real rate exhibited small fluctuations around an average of 6.5%, a level much higher than the values reached in the 1970s. This pattern resulted from the gradual decline in nominal rates following the decline of the inflation rate, and reflected the policy makers' concern with preventing tensions on the balance of payments due to excessive growth of aggregate demand.

These events are reflected by the estimates reported in Table 2. The reaction function, estimated from 1979.4, is specified as proposed in GUISO [8] for the discount rate. The average nominal rate on Treasury bills displays a slow adjustment to the expected inflation rate of consumer prices, as described by the expectations obtained from the *Mondo economico* surveys (VISCO [14]), which are endogenized in the model. The increases in the nominal rate that were aimed at contrasting aggregate demand shocks that could cause balance of payments tensions are captured by the negative effect of the ratio of the current account balance to total imports. Formally:

$$(2) \quad i = (1-\beta_1)\,\bar{r} + (1-\beta_1)\,\beta_2\,q^e + \beta_1\,i_{-1} - (1-\beta_1)\,\beta_3\,s_{-1}$$

where i is the short-term Treasury bills rate, q^e is the expected rate of inflation, and s is the ratio of the current account balance to imports. The parameter $\bar{r}$ can be interpreted as the real interest rate monetary authorities aim to maintain. The rule, characterized parametrically by the values of β_1 and $\bar{r}$, implies, for $\beta_2 = 1$, that the policy makers, having chosen the target real rate, slowly adjust the nominal rate in response to expected changes in inflation and raise it when facing current account deficits.

The estimated coefficients of the interest rate rule appear very stable and satisfactorily capture its actual evolution, implying an equilibrium real rate of approximately 6%. In the regressions over sample periods including 1985 and 1986, we introduced a dummy variable for 1985.1, when the nominal rate underwent a sudden decline that did not continue in the following quarters (Graph 3).

Concerning the predictive stability of the estimated regression, it should be noticed that the value for the *Chow* test statistic is not negligible when comparing the estimates from the period ending in 1985.4 (used in the simulations) with those for the period ending in 1986.4. This essentially reflects the reduction in size and the loss of significance of the intercept. This result indicates an important fact. With the reduction in the imported component of inflation, owing to

TABLE 2

REACTION FUNCTION OF THE NOMINAL INTEREST RATE OF TREASURY BILLS (*)

T	1983.4	1984.4	1985.4	1986.4	1985.4	1986.4	1986.4
c_0	1.962	2.092	1.155	0.311	1.610	0.418	1.745
	(.571)	(1.070)	(1.174)	(0.418)	(1.658)	(0.565)	(2.084)
c_1	0.725	0.705	0.739	0.772	0.725	0.771	0.719
	(5.040)	(7.207)	(11.872)	(13.509)	(12.140)	(13.644)	(13.178)
c_2	0.228	0.239	0.268	0.291	0.251	0.284	0.250
	(2.255)	(4.432)	(6.507)	(7.462)	(6.234)	(7.295)	(6.702)
c_3	−6.893	−8.193	−7.088	−6.628	−7.320	−6.631	−7.193
	(1.367)	(2.005)	(2.356)	(2.186)	(2.554)	(2.211)	(2.670)
c_4	−1.050	−1.063	−1.090	−1.115	−1.075	−1.107	−1.087
	(2.508)	(2.332)	(2.334)	(2.321)	(2.419)	(2.329)	(2.554)
c_5					−1.017	−.738	−1.038
					(1.755)	(1.234)	(1.895)
c_6							−1.038
							(2.589)
R^2	0.932	0.945	−0.953	−0.968	0.959	0.970	0.977
SER	0.613	0.554	0.568	0.585	0.540	0.579	0.518
DW	2.547	2.512	2.693	2.633	2.447	2.411	2.492
MLM1	1.674	1.543	3.221	2.931	1.536	1.279	1.344
MLM4	0.247	0.229	0.571	0.002	0.146	0.003	0.947
ARCH1	0.352	1.578	0.612	0.002	1.134	0.341	1.197
N	17	21	25	29	25	29	29
CHOW	0.264	0.222	0.377	0.480			
CHOW1						1.839	

(*) Estimated equation:
$TABOT = c_0 + c_1 \, TABOT_{-1} + c_2 \, INFEY + c_3 \, \Sigma_{i=1}^{4}(BALCURD/IMPD)_{-i} + c_4(DU802 + DU803) + c_5 \, DU851 + c_6(DU862 + DU863 + DU864)$.
Estimation period: 1979.4 − T.
Variables: BALCURD = current account balance (seasonally adjusted); DUXYZ = dummy (= 1 in 19XY.Z); IMPD = imports of goods and services (seasonally adjusted); INFEY = annual inflation rate of consumer prices, expected over the following semester; TABOT = interest rate on short term Treasury bills.
Legend: (See TABLE 1): CHOW1 = test of predictive stability, performed with respect to estimates over a sample period ending in 1985.4.
Source: BANCA D'ITALIA, DATA BANK Quarterly Econometric Model (as of April 1987).

the fall of oil prices and the consequent improvement in the Italian balance of payments, since the Spring of 1986 monetary policy started to reduce its degree of stringency. In conditional simulation exercises, not only the assumption of invariance of policy variables at historical values but also the assumption of invariance of policy rules require caution. In fact, it is likely that rules would have been different if the constraints had assumed values very different from those prevailing when the rules were put in place.

Even though stylized, the two reaction functions are analogous to similar recent attempts (See for instance, BASEVI, CALZOLARI, and COLOMBO [2]). They aim at taking into account the links between exchange rate and monetary policies, on the one hand, and inflation and the balance of payments, on the other. They are specified in a flexible form that allows the simulation of behaviors different from those experienced in history. For instance, one can obtain a rule maintaining the real exchange rate at a given level by setting the coefficient of relative prices equal to 1 and all other coefficients equal to 0 and by changing the value of the intercept term. A similar procedure can be easily followed for the Treasury bills rate, to simulate the effects of a less gradual or more restrictive monetary policy.

Although they do not presume to exhaustively describe the reaction of monetary authorities when faced with all the diverse shocks of the 1980s, these reaction functions describe well the actual evolution of the variables under investigation. It is beyond the scope of this paper, even if interesting, to consider functions estimated in terms of monetary and financial aggregates rather than in terms of exchange and interest rates.

In Section 4, an attempt is also conducted to determine what the outcome would have been if the second oil shock (and the recent negative shock) or the appreciation (and recent depreciation) of the dollar had not occurred. It was then necessary also to endogenize the most important international variables, in order to allow for fluctuations in oil prices and the dollar exchange rate (measured here relative to the deutsche mark) to affect world demand and the international prices of manufactured goods and raw materials. Details on the relevant estimated equations are presented in the Appendix.

Finally, some variables that are exogenous in the current version of the econometric model were endogenized ad hoc. These are some minor items in the public sector's balance sheet, most of which have been set to change proportionally to nominal GDP or some other endogenous variable. Fiscal policy instruments (tax rates, employment in the government sector, public consumption, and investment in 1970 prices) have been kept unchanged at their actual levels. This amounts to assuming strong exogeneity, an assumption to be kept in mind when examining some of the experiments reported below (6). For good and services subject to price control, including energy products, their relative price was kept constant. As for capital movements other than trade-related credit flows, they were kept exogenous at their historically observed values, given the existence of extensive capital controls in the priod (7).

3.2 *The Price-Wage Sector of the Econometric Model*

In order to illustrate the main transmission mechanisms, it is useful to summarize the price-wage block of the model. The essence of its structure can be described by the following basic relations [see Banca d'Italia [1] for further details].

A first equation defines the growth rate of nominal wages in manufacturing. Defining with lower-case letters natural logarithms and with Δ the first-difference operator, the estimated equation amounts to

$$(3)\ \Delta w_t = \alpha + \sum_{i=1}^{4} \omega_1 u_{t-1} + \beta_1 \eta_t \Delta c_{t-1} + \beta_2 (1 - \eta_{t-1})\ _{t-2}\Delta c^e_{t-1}$$

$$+ \beta_3 (1 - \eta_{t-1}) [\Delta c_{t-2} -\ _{t-3}\Delta c^e_{t-2}]$$

---

(6) However, for many of these variables the assumption of strong exogeneity does not seem dangerous as they change very slowly over time and do not react systematically to cyclical conditions.

(7) In the capital accounts of the balance of payments, only short-term import/export financing is endogenous, while other capital movements are exogenous (and thus kept equal, in the simulations, to their historical level). This could lead, in some simulations, to underestimate exchange rate movements, given the reaction functions we adopted.

where w, c, and u are, respectively, the hourly wage rate in manufacturing, the *scala mobile* cost-of-living index, and the rate of unemployment (cyclically adjusted to account for employees on wage supplementation). $_{t-i-1}\Delta\, c^e_{t-i}$ defines the rate of inflation expected at $t-i-1$ for quarter $t-i$ and is derived from *Mondo economico* surveys (see Visco [14]). η is an endogenous variable that defines, for each quarter, the degree of indexation of wages to the *scala mobile* index (for details, see Visco [15]). After statistical testing, the β_i coefficients have been restricted to unity.

This equation amounts to an expectations-augmented Phillips curve, modified for the presence of indexation. It includes also a catch-up term, defined so that previous forecast errors appear with a coefficient of 1 minus the actual degree of indexation. This implies that the actual inflation rate has a permanent effect on the wage equation, while expectations have only a transitory effect. Wages in other sectors are linked to the wage rate in manufacturing by means of homogeneous relations. This reflects the hierarchical nature of wage negotiation in Italy, recognizing a leading role to manufacturing. Given the lag structure of the wage equations for the private sector, an inflation shock is reflected by wage rates for two-thirds in the first two quarters, and fully in two years.

Nominal wages react with some lag to changes in unemployment, as can be seen from equation *(3)*. According to the model estimates, a 10% increase in the unemployment rate leads to a reduction in the annual percentage rate of change of nominal wages equal to 0.5% after two quarters, and to a total 1.2% reduction after five quarters.

A second important relation determines the value-added deflator in manufacturing. Formally:

$$(4) \qquad p^v_t = \mu_t + \beta\,(w_t + h_t - \pi_t) + (1 - \beta)\, p^v_{t-1}$$

wher p^v and π are, respectively, the logarithms of the value-added deflator and the productivity trend, h is the logarithm of (1 plus) the rate of social insurance contributions, and μ is a function of short-run cyclical changes in the mark-up on unit labor costs. In particular, μ depends on the rate of capacity utilization (with an equilibrium elasticity of 0.1), and the competitive position of Italian firms. In the very short term, a sudden acceleration in input prices is also reflected in a temporary reduction of the mark-up, and viceversa. The estimate of β is slightly over one-third, implying a somewhat slow adjustment of prices to changes in labor costs — which eventually are fully reflected in the deflator. Similar specifications have been adopted in modeling value-added deflators in the energy, construction, and service sectors. In agriculture, the institutional features of the European Common Market have been imposed.

The price of domestically produced goods is defined as a weighted average of the value-added deflator in manufacturing and the prices of imported and domestic inputs. Demand deflators are linked to value-added and import deflators by means of composition equations. The specification of the consumption deflator — to which the *scala mobile* index is also linked — attempts to account for the dynamic pass-through from wholesale to consumer prices. The consumption deflator, net of energy and other regulated prices, dynamically and fully adjusts to a composite price index which includes domestically produced and imported final goods. Two-thirds of the adjustment take place in about two quarters. Finally, a detailed set of equations models energy prices and other prices subject to government control, accounting for the most important institutional aspects of their determination.

The most important variables which affect consumer prices, exercising inflationary or deflationary impulses, are then: foreign prices, through the direct and indirect effects of higher import prices; inflationary expectations, through their effect on labor costs; publicly regulated prices; economic conditions, expressed by the unemployment rate and the rate of capacity utilization; and the exchange rate. Essentially, the effects produced by the exchange rate are: *a)* the direct effect on the price of imported final goods denominated in lire, *b)* the indirect effect through the higher cost of imported intermediate inputs, *c)* the effect on economic activity, to the extent to which a nominal depreciation affects competitiveness. These effects are sustained by the fact that wages are fully indexed to prices in less than a year, through the *scala mobile* indexation mechanism and the catch-up term specified in equation *(3)*.

It is important to notice that the exchange rate rule used in the simulations and described in Section 3.1 implies accommodating price changes of imported intermediate inputs and reacting negatively to price changes of imported final goods. This follows from the fact that the policy rule is defined in terms of competitiveness of final goods while foreign inputs enter the production of domestic manufactured goods.

3.3 *The Simulations*

The simulations presented in the following pages are compared to "history" (8). Technically, this means that static simulations have been conducted, computing residuals so as to replicate history, and keeping the residuals constant at their historical values (treating them like intercept terms) in the conditional simulations, while modifying the pattern of the relevant exogenous variables or the parameter values of individual equations.

The simulation exercises aim in the first place at identifying the contribution of international variables to inflation, in order to measure its imported component; they aim also at evaluating separately the effects of oil prices and the exchange rate of the dollar. Three experiments have been conducted with these goals:

a) SIM 1: keeping the dollar price of imported energy products at its 1979.1 level.

---

(8) The model has been estimated with the data available in October 1986. Thus, in the estimation step, we could take into account neither of the two major NIA revisions that were subsequently released: the revisions of ISTAT Quarterly Accounts (ISTAT, 1986) and the revisions in the annual series presented in *Relazione generale sulla situazione economica del paese* for 1986. The simulations use the data available by December 1986, which incorporate the mentioned revisions in the quarterly data, but not those in the annual series. A realignment of all equations in the model was, therefore, necessary. Static simulations were run for each equation, using the coefficients estimated with the old data and the "new", revised time series. The residuals are thus to be imputed not only to the estimation procedures but also to the revisions in the data. In order to evaluate the relevance of these revisions, we ran several simulations within the sample (among others, those presented in BANCA D'ITALIA [1]), using the "old" and the "new" data. We consistently obtained very similar values for the multipliers. The properties of the model do not appear to be altered by the use of the new quarterly series. We chose then to use the new quarterly series as they allow one to extend the analysis to all of 1986 and, thus, to evaluate the effects produced by the negative oil shock.

b) *SIM* 2: keeping the dollar price of imported energy products and the dollar/deutsche mark exchange rate at their 1979.1 levels.

c) *SIM* 3: keeping all international prices and the dollar/deutsche mark rate at their 1979.1 levels.

In these exercises, some international variables which are normally exogenous to the model (international prices of agricultural and non-agricultural raw materials, export prices of manufactured goods of Italy's competitors, world demand) have been endogenized as discussed in the Appendix.

In the model, before-tax prices of energy products, such as electricity and heating oil, are linked to the output price of refined oil products through the corresponding historically observed ratios. These ratios changed sensibly in the period, due to the regulated prices policy followed in response to the evolution of oil prices. Thus, the ratios too have been kept constant at the values observed in 1979.1, in order to keep the price of these energy products unchanged, relative to the price of refined oil products.

Moreover, in *SIM* 2 and *SIM* 3, wher the dollar/deutsche mark rate is kept constant at its 1979.1 level, the exchange rates between the deutsche mark and all other currencies, except the lira (whose effective exchange rate is determined by the reaction function previously discussed), are kept at the values observed in 1979.1. Therefore, the exchange rates between these currencies and the dollar are determined consistently by the simulated evolution of the dollar/deutsche mark rate. This amounts to making the extreme assumption that the latter does not affect the exchange rate between the deutsche mark and the other currencies (i.e., the fluctuations observed in the latter rates have been determined not by differentiated effects of the dollar but rather by causes specific to the various countries (9)).

After obtaining a quantitative estimate of the domestic component of inflation, the effects produced by the policies of the

---

(9) Although this is an extreme assumption not necessarily supported by theoretical analysis or empirical evidence, nevertheless, it is likely that, for the goal of this paper, neglecting the effect of changes in the dollar/deutsche mark rate on the exchange rate structure within the European Monetary Union is of second-order importance.

period have been evaluated. In the experiments, the international variables have been kept at their historical values. In experiment *SIM* 4, the effects have been considered resulting from the regulated prices policy followed since the second half of 1981 by simulating an inflation-neutral policy. In the model, the regulated prices of nonenergy products are endogenized as functions of consumer prices, and the regulated prices of energy products as functions of the output price of refined oil products, keeping the ratios between these prices constant and letting the absolute prices grow as consumer prices and as the output price of refined oil products grow.

As far as exchange rate policy and monetary policy are concerned, they were treated, as discussed in Section 3.1, not as a series of occasional measures but as behavioral rules systematically followed by the policy makers over the period. This allows one to measure their contributions — and their costs — by contrasting them with alternative rules. Four exercises of this type were conducted:

a) *SIM* 5: the exchange rate reaction function has been changed into an extreme, fully accommodating rule: i.e., a rule that maintains the real exchange rate at a given level (1979.1) (10).

b) *SIM* 6: the exchange rate rule is nonaccommodating: i.e., the nominal exchange rate is kept at its 1979.1 level.

c) *SIM* 7: the interest rate reaction function has been changed so that it reacts to expected inflation only, keeping the real interest rate at its 1981.1 level (which is the rate prevailing, on average, before the second oil shock).

d) *SIM* 8: in this experiment, in order to measure the costs of the "gradualness" of the interest rate policy of the period, the real interest rate is raised to 7.6% (its 1981.4 level) in 1980.1, instead of reaching that value gradually, and is subsequently kept equal to history.

Finally, in experiment *SIM* 9, the contribution to disinflation given by the precommitment policy is evaluated. To measure its

---

(10) Agents are assumed to know the exchange rate reaction function and to form their expectations consistently. Thus, in the model, the expected change in the exchange rate is obtained from the expected values of the regressors (generated by *ARIMA* processes) entering the reaction function, given its estimated coefficients. After changing the parameters of the policy rule, the parameters of the equation that generates exchange rate expectations have also been accordingly modified.

effects, including those generated by the drop in inflationary expectations following the announcement of a target inflation rate, the evolution of the economy has been simulated after eliminating those policy measures. In particular: 1) the value of the indexation provisions was restored to its level prior to 1983; 2) the *scala mobile* cost-of-living index was raised by the 4 points artificially removed during the first semester of 1984; 3) the growth rate of publicly regulated prices was set equal to the inflation rate of consumer prices since 1983.4; 4) the effects on rents caused by the deferment of the "fair rent" indexation in 1984 were eliminated (11) and 5) the forecast error (always positive in the relevant period) of the equation explaining inflationary expectations on consumer prices was imputed to the effect produced by the announcement of a target inflation rate, and set equal to zero in the exercise.

The most important results of these exercises are described in Tables 3, 4, 5, and are discussed in the next section. Detailed results are available from the authors upon request.

4. - Disinflation: Interpreting the Results

Although the reduction of the inflation rate between 1980 and 1986 looks like a homogeneous process, several different events operated on its specific time pattern. In order to determine what has been quantitatively relevant, the simulations are intended to measure

---

(11) The adjustment of rents to cost-of-living increases for 1984 was suspended by the government in July 1984. In order to provide for an appropriate evaluation of the contribution given by this measure to curbing inflation, we assumed a normal adjustment of rents to changes in the cost-of-living index for 1984, i.e., an increase equal to 75% of the annual change of the index (the indexation allowance mandated by law). In the exercise, we first evaluated the increase in the index of rents which would have otherwise taken place, reintroducing the indexation allowance. After computing the difference between the rate of growth of the "simulated" and the "actual" indices for the period of 1984.10-1985.9, we raised the inflation rate of rents, as defined by the National Income Accounts, by 50% of that difference. This coefficient was intended to take into account the different composition of the consumer price index of rents relative to the National Accounts index (which includes imputed rents of owner-occupied houses) as well as the fact that the former also includes living arrangements outside the regulated regime, unaffected by the suspension of the adjustment.

TABLE 3

INTERNATIONAL VARIABLES
(average annual percentage changes)

Year	History	SIM 1	SIM 2	SIM 3

Import price of nonagricultural raw materials (in US$)

1979	41.9	29.3	28.4	20.4
1980	29.2	13.7	13.2	0.0
1981	−18.3	−16.1	−6.0	0.0
1982	−13.1	−8.1	−2.4	0.0
1983	2.3	6.8	11.6	0.0
1984	−0.4	0.0	6.7	0.0
1985	−4.5	−5.3	−2.6	0.0
1986	8.4	33.2	14.2	0.0

Import price of agricultural raw materials (in US$)

1979	12.4	5.0	4.4	3.2
1980	8.0	−2.9	−3.3	0.0
1981	−8.0	−8.3	1.3	0.0
1982	−7.1	−4.3	−0.1	0.0
1983	−3.4	0.0	3.1	0.0
1984	3.0	4.1	9.7	0.0
1985	−4.1	−4.2	−2.4	0.0
1986	21.0	41.8	23.4	0.0

Export price of manufactured goods, Italy's competitors (in national currencies)

1979	8.1	7.3	7.4	4.2
1980	8.7	5.4	5.7	0.0
1981	8.3	7.6	4.1	0.0
1982	8.0	8.7	7.3	0.0
1983	3.2	4.0	3.4	0.0
1984	4.2	4.7	3.0	0.0
1985	3.8	3.8	2.9	0.0
1986	−1.6	0.5	5.5	0.0

World demand (in US$)

1979	5.3	5.3	5.3	5.3
1980	0.7	4.9	4.7	4.1
1981	1.2	8.9	8.9	7.7
1982	−1.0	1.1	3.9	3.0
1983	1.9	0.8	2.0	0.7
1984	8.7	6.2	7.0	6.2
1985	4.1	2.9	4.4	3.9
1986	3.8	3.2	3.4	2.9

TABLE 4

SIMULATION RESULTS: *SIM* 1-*SIM* 4

Year	History	*SIM* 1	*SIM* 2	*SIM* 3	*SIM* 4
Consumer price index (average annual percentage change)					
1979	14.8	13.5	13.5	13.2	14.8
1980	21.2	15.5	15.8	16.6	21.2
1981	19.6	15.5	13.8	13.6	19.5
1982	16.5	14.3	12.5	10.4	16.3
1983	14.7	12.5	11.7	8.7	13.8
1984	10.8	8.7	7.5	4.2	10.4
1985	9.2	7.5	6.2	2.8	8.9
1986	5.9	8.6	10.5	6.2	4.5
GDP (average annual percentage change)					
1979	4.9	5.0	5.0	4.6	4.9
1980	3.9	5.0	5.1	3.5	3.9
1981	0.2	4.1	3.4	2.2	0.2
1982	− 0.5	1.6	2.0	1.4	− 0.5
1983	− 0.2	− 0.5	0.1	0.5	0.0
1984	2.8	1.8	1.4	2.3	3.0
1985	2.3	1.5	2.1	2.5	2.4
1986	2.8	2.1	3.5	3.6	3.1
Current account balance (billions of lire)					
1979	4542	6845	6690	6842	4552
1980	− 8291	820	741	1386	− 8291
1981	− 9225	5302	7484	7344	− 9218
1982	− 7412	5366	7332	7053	− 7493
1983	1183	10280	11490	9557	650
1984	− 5084	2322	4650	3648	− 5812
1985	− 8032	− 1456	768	− 1356	− 9001
1986	6792	− 401	− 3945	− 5731	4792

the importance of the various events that occurred in the period. In this section, using the results of these experiments, an attempt is made to provide an interpretation of the disinflation process in terms of the international factors and the economic policies at work.

TABLE 5

SIMULATION RESULTS: SIM 5-SIM 9

Year	History	SIM 5	SIM 6	SIM 7	SIM 8	SIM 9
Consumer price index (average annual percentage change)							
1979	14.8	15.6	14.8	14.8	14.8	14.8
1980	21.2	25.1	20.1	21.2	21.1	21.2
1981	19.6	22.0	15.4	19.5	19.0	19.5
1982	16.5	18.8	11.1	16.7	16.0	16.4
1983	14.7	20.5	10.4	15.5	14.3	14.7
1984	10.8	17.2	6.3	11.6	10.3	11.3
1985	9.2	14.3	4.4	10.2	8.7	9.9
1986	5.9	10.4	2.8	7.6	5.5	6.1
GDP (average annual percentage change)							
1979	4.9	5.2	4.9	4.9	4.9	4.9
1980	3.9	5.0	3.5	3.9	3.5	3.9
1981	0.2	−0.4	−0.6	0.2	−0.4	0.2
1982	−0.5	−0.8	−0.6	0.1	0.2	−0.5
1983	−0.2	0.7	0.3	0.5	−0.1	−0.2
1984	2.8	3.0	2.9	2.6	2.9	2.8
1985	2.3	1.6	2.5	2.3	2.5	2.1
1986	2.8	3.2	3.2	3.0	2.8	2.5
Current account balance (billions of lire)							
1979	4552	3796	4634	4552	4552	4552
1980	−8291	−10206	−7217	−8291	−7325	−8291
1981	−9225	−8871	−7132	−9361	−7104	−9225
1982	−7412	−7428	−7444	−9500	−6360	−7412
1983	1183	2810	−1530	−2468	2208	1177
1984	−5084	−2302	−8113	−9209	−4701	−5172
1985	−8032	−772	−10752	−13380	−8160	−7369
1986	6792	19592	−2534	2538	5981	8177

4.1 *The Internal and External Components of Inflation*

In an open economy with a managed exchange rate regime, changes in foreign prices inevitably affect domestic prices. Moreover, the more open the economy, the larger their impact. Thus, the first step has been at measuring the contribution of international variables to inflation in Italy and also at evaluating separately the effects of oil prices and of the dollar exchange rate (simulations *SIM* 1, *SIM* 2, and *SIM* 3). Table 3 reports the rates of growth of the international variables most relevant for these experiments.

The main results are summarized in Table 4. According to experiment *SIM* 1, had the import price of energy products remained unchanged since 1979.1, the inflation rate would have been lower than the historical rate, except in 1986. Keeping the import price of energy products constant at its 1979.1 level amounts to eliminating the second oil shock, its small but continuous decline between 1982 and 1985, and the negative shock of 1986. Therefore, relative to history, the growth rate of international prices is lower until 1981 and higher later, while the growth rate of world demand is higher until 1982.

As can be seen from the simulation results, this variable is particularly important in 1980. Had the second oil shock not occurred, the inflation rate would have been 5.7 points lower. In 1986, on the contrary, the inflation rate would have been 2.8 points higher. The effect of the simulated evolution of the price of energy products on inflation is strengthened by the exchange rate reaction, given the policy rule followed. In fact, due to a lower inflation differential relative to the other countries in the first half of the period and higher foreign currency reserves, the effective exchange rate devalues in the simulation by less than in history and is 21% lower (more appreciated) by the end of the period.

In experiment *SIM* 2, the joint contribution given by the two most important (and, according to our assumptions, independent) international exogenous variables is evaluated. Historically the dollar has undergone a large appreciation relative to the deutsche mark between mid-1980 and early 1985, and a rapid devaluation afterwards. In this experiment, the rate of inflation of consumer prices shows a pattern similar to the previous simulation, resulting lower that the observed rate until 1985. Moreover, the dollar appreciation that occurred after 1981 appears to have strengthened the effect of the second oil shock on inflation. In fact, in this exercise the difference between actual and simulated inflation is larger than in the previous experiment.

Finally, in *SIM* 3 a quantitative estimate is obtained of the contribution of foreign factors to inflation, by setting the rate of change of all foreign prices equal to zero and by keeping the exchange rate between the dollar and deutsche mark constant. Given the wide fluctuations experienced by these variables in the period under consideration (Table 3), the shocks whose effects are being evaluated are very strong. In the light of considerations concerning the stability of stochastic equations in the presence of large changes in the exogenous variables, the results should however be evaluated with caution.

From this simulation (Graph 4), estimates are obtained of the overall contribution of intenational factors and of "endogenous" inflation, i.e., the rate of inflation net of any direct and indirect effects generated by changes in foreign prices (12). In each period, the measure of the endogenous component depends also on the initial conditions (i.e., the rate of inflation inherited from the past at the starting point of the simulation) and on the speed of adjustment of the system. This problem does not affect the estimate of external inflation, which is independent of initial conditions, being obtained as the residual (13). In order to eliminate from the estimate of endogenous inflation the effect of the initial conditions, this has been estimated by means of a simulation of the price-wage block conducted with the values of both endogenous and exogenous variables, prior to 1978.4,

---

(12) The experiment is conducted with the simulated evolution of international prices and the dollar/deutsche mark exchange rate affecting world demand, as discussed in the Appendix. As the size of the elasticity of world demand with respect to the price of oil is controversial and the estimate obtained in this paper appears higher than others, it is necessary to interpret cautiously the impact of a similar shock. The same experiment was then conducted keeping world demand at its actual historical values instead of endogenizing it. Even if the rate of inflation is now substantially higher than the historical one, the results (Graph 4) confirm the finding that the time pattern of inflation was not affected, except in 1986, by the contribution of international variables. The evolution of real variables is, however, very different in the two exercises: When world demand is exogenous at its historical values, not only do we observe a fall in GDP relative to the simulation with endogenous world demand but also, for the first four years, relative to history. Similarly, the unemployment rate is higher compared to its actual level over the entire period.

(13) This is exactly true only when the model consists of a system of linear difference equations.

GRAPH 4

SIMULATION RESULTS FOR CONSUMER PRICE CHANGES AND UNEMPLOYMENT (*SIM* 3 SIMULATIONS) (*)

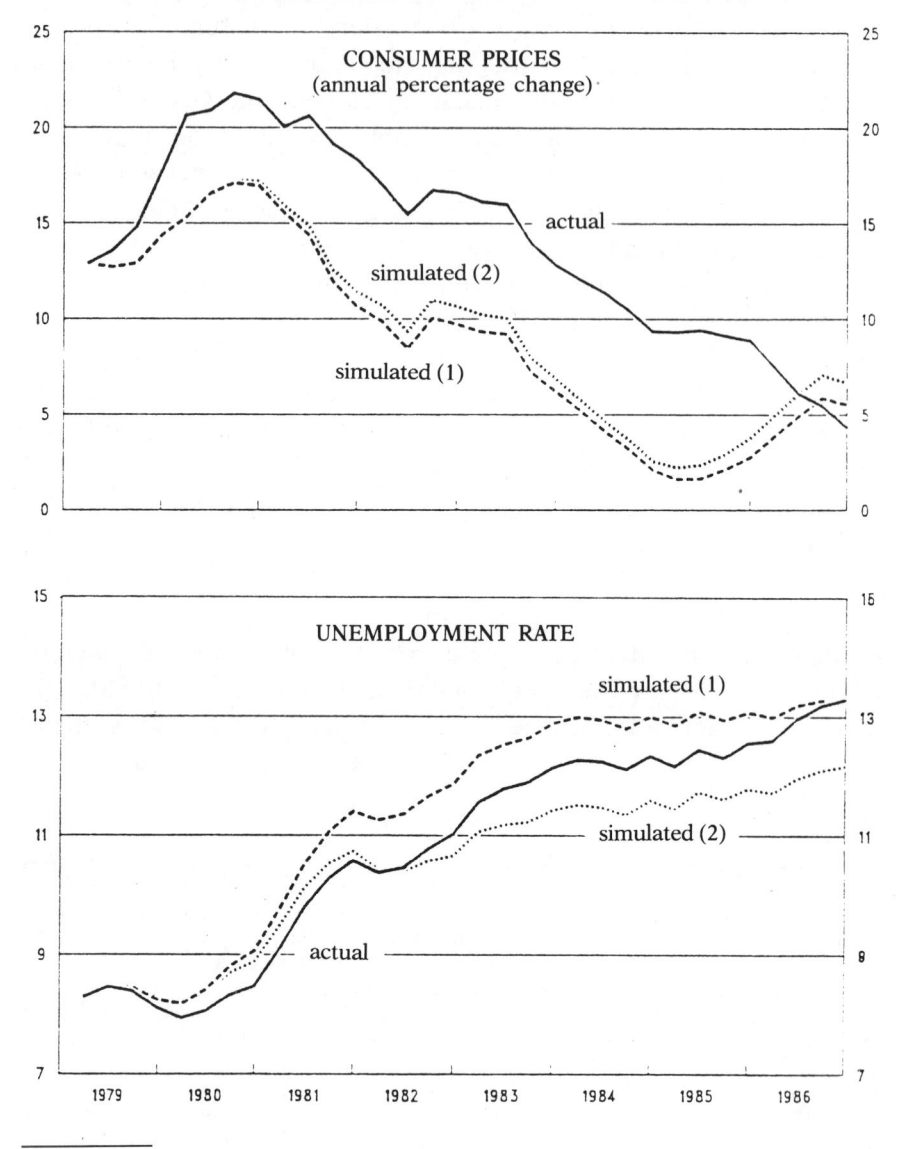

(*) Simulations with international prices and the dollar/deutsche mark rate constant at their 1979.1 levels, and world demand exogenous (1) and endogenous (2), respectively. The unemployment rate is corrected for workers temporarily in wage supplementation (*Cassa integrazione guadagni*).

set equal to their levels in that quarter, and the values of all variables defined as rates of growth (e.g., inflationary expectations) set equal to zero until 1978.4.

Table 6 shows the decomposition of inflation obtained in this way for the period 1979-1986, distinguishing between inertial, endogenous, and foreign components. First of all, one notices the magnitudes of the impulses originated abroad, which contributed to a remarkable extent to the inflation rate in Italy, except in 1986. The endogenous component is nevertheless predominant until 1983, and declines significantly during the two following years; the inflation rate in 1986, instead, is entirely determined by domestic factors. As one would anticipate, the effect of initial conditions is important in the first year of the simulation and then loses relevance quickly.

On the basis of the above decomposition, it is possible to compute the contribution given to the time pattern of inflation by international prices and that of internal origin, due to the effect of policies or changes, independent of policies, in domestic variables.

In each period, the observed inflation rate was expressed as

$$(5) \qquad q_t = q_t^I + q_t^{EN} + q_t^*,$$

where q^{EN} is the measure of endogenous inflation in Table 6, q^* is the component that can be attributed to the foreign sector, and q^I

TABLE 6

DECOMPOSITION OF THE INFLATION RATE

Year	History a	Inertial b	Endogenous c	Imported d
1978	12.1			
1979	14.8	5.3	7.9	1.6
1980	21.2	1.3	15.5	4.6
1981	19.6	1.1	12.5	6.0
1982	16.5	0.9	9.5	6.1
1983	14.7	1.1	7.6	6.0
1984	10.8	0.7	3.5	6.6
1985	9.2	0.3	2.5	6.4
1986	5.9	0.1	6.1	−0.3

TABLE 7

CHANGE OF THE INFLATION RATE

Year	Change in historical inflation	Change due to: inertial component	Change due to: endogenous component	Change due to: imported component
1979	2.7			
1980	6.4	−4.0	7.6	3.0
1981	−1.6	−0.2	−3.0	1.4
1982	−3.1	−0.2	−3.0	0.1
1983	−1.8	0.2	−1.9	−0.1
1984	−3.9	−0.4	−4.1	0.6
1985	−1.6	−0.4	−1.0	−0.2
1986	−3.3	−0.2	3.6	−6.7

represents the effect of initial conditions. Taking first differences, the variation of the rate of inflation ($q_t - q_{t-1}$) can be decomposed in the endogenous contribution ($q_t^{EN} - q_{t-1}^{EN}$), the foreign contribution ($q_t^* - q_{t-1}^*$), and the inertia effect ($q^I - q_{t-1}^I$). These are presented in Table 7. The main result is that the process of disinflation has been dominated, with the exception of 1986, by the decline of the endogenous component. This contrasts several common theses, which assert that the decline in inflation between 1980 and 1985 was due mainly to the weakening of the external contribution.

Before examining the role played by exchange rate and monetary policies, which we believe are the most important determinants of the disinflation process, and some other relevant events such as the precommitment experiment, it may be useful to comment further on Table 7. The surge of inflation in 1979-1980 was fed by internal impulses, mainly strong excess demand in 1979, and worsened by the second oil shock, which was at least partially accommodated (14). The decline of the inflation rate in 1981 and 1982 was due to the reduction of the endogenous component, caused presumably by the effects of the partial exchange rate accommodation and the restrictive

---

(14) The reader is reminded that the exchange rate rule implies that the policy makers accommodate shocks on the price of imported inputs.

monetary policy implemented between the end of 1979 and the beginning of 1982. In 1983 the endogenous contribution to the disinflation process came temporarily to an end, owing in part to the strong increase in publicly regulated prices (*SIM* 4; See Table 4). In addition, value-added tax rates rose between 1982 and 1983, following the increase in public expenditure, mostly in 1982. After removing these impulses, the rate of inflation in 1983 is about 1 point lower (15). The largest reduction, occurring in 1984, takes place at the same time as the precommitment agreement (an issue on which we shall return later), and is entirely caused by the reduction of endogenous inflation, while the contribution of foreign prices is positive and thus contrasts this tendency. Internal inflation lessens its reduction considerably in 1985 and even starts increasing again in 1986. Had the strong deflationary effect produced by the collapse of oil prices, which reduced the inflation rate 6.8 points, not taken place, the overall rate of inflation would have increased in 1986.

4.2 *The Policies and Their Role*

The behavior of monetary authorities has been synthetically described by the reaction functions of the exchange rate and the interest rate presented in previous sections.

In the period between 1980 and 1986, the monetary authorities have behaved homogeneously in response to diverse events. For the exchange rate, such behavior was exhibited by a slow (and empirically partial) adjustment of the nominal rate to the price differential. For the interest rate, having chosen the value of the real rate, the rule implies a slow adjustment of the nominal rate to expected inflation. Moreover, unfavorable developments in the current account are contrasted with an increase of the interest rate; excessive losses of foreign reserves are countered with exchange rate depreciations.

The effects of these policies, their relative effectiveness, and their

---

(15) In order to evaluate the effect of these impulses, we conducted an experiment in which the value-added tax rates are kept constant at their level prior to the rise which occurred in 1982.3, and the ratio between public sector consumption and GDP is kept at its 1981.1 value.

costs, in terms of attaining a given rate of inflation, have been evaluated by varying the parameters of the rules. A few qualifications should however be advanced. First of all, the alternative rules to which the policies actually followed are compared must be sustainable. "Sustainability" is here defined in terms of the value of the ratio of reserves to imports. Historically this ratio reached a minimum value of 10% during the exchange rate crisis of 1976. An alternative policy rule is defined as sustainable if, in the resulting simulation, that ratio is not less than 20%, as it is reasonable to assume that rules would start to change in proximity of a "danger level".

Secondly, the issue of the attainability of the rules has not been addressed. This problem is relevant for the exchange rate rule as, for instance, when a perfectly accommodating exchange rate policy has been assumed. In fact, this might have implied an exchange rate not always compatible with the fluctuation bands allowed by the European Monetary Agreement. In this case, the assumption is made that if the exchange rate had exceeded those thresholds, then a parity realignment would have taken place.

Thirdly, it has been assumed that changes in the parameters of the rules would not affect the structural parameters.

Finally, it must be emphasized that the parameters of the two rules are, in general, not independent, as they originate from the same policy makers. However, since they have not been derived from the solution of a hypothetical optimization problem, it cannot be established how the parameters of one rule should change when changes in the parameters of the rule are simulated.

4.2.1 The Role of the Exchange Rate

To obtain an estimate of the importance of the exchange rate regime, in terms of evolution of the inflation rate, two extreme cases have been examined, where the authorities follow respectively a policy in which the real exchange rate is fixed at a given level and a policy in which the nominal exchange rate is kept constant.

The exchange rate rule adopted in experiment *SIM* 5 fully and instantaneously accommodates the inflation differential between the domestic and foreign prices of manufactured goods (denominated in lire). That is, the rule is imposed:

(6) $$e_t = p_t - p_t^* + \alpha_0,$$

where $\alpha_0 = e + p^* - p$ in 1979.1. Qualitatively this extreme experiment shows how the exchange rate policy has been indeed important in bringing inflation under control after the second oil shock. Quantitatively (Table 5), inflation would have been 4 points higher, on average, per year between 1979 and 1986.

It is apparent, however, that the policy actually followed has not been costfree. Because of the more favorable evolution of competitiveness (in terms of final rather than manufactured goods, for which it is constant by assumption), the fully accommodating rule implies higher values for GDP as well as for the balance of payments. In fact, there would have been a gain of 6,600 billion between 1979 and 1986, in terms of GDP in 1970 lire, equal to 0.8% of cumulated GDP in the same period. At the same time, there would have been larger positive (or smaller negative) balances in the current account, except for 1979 and 1980, when the oil shock took place.

It is beyond the scope of this paper to establish which of the two regimes (the actual or the simulated) would have been "better". This would require explicit welfare functions of the policy makers. However, concerning GDP alone the gains following the adoption of the accommodating rule appear rather small.

The exchange rate rule has been contrasted also with a nonaccommodating reaction function, according to which the nominal exchange rate does not respond to changes in relative prices and is fixed at its 1979.1 level (*SIM* 6). From this experiment as well, the high responsiveness of domestic prices to the behavior of the exchange rate is apparent. Adopting this nonaccommodating rule would have allowed sizable gains in terms of reducing inflation. Its rate would have on average been over 3 percentage points lower than history.

As a counterpart to these gains in terms of inflation, this policy would have caused a decline in GDP. A synthetic index of the cost implied by a disinflationary policy is the sacrifice ratio, i.e., the cumulated percentage of GDP following a permanent drop of 1 percent in the rate of inflation in a given period. This measure, which is not intended to have welfare implications, is equal to 6.6 in the experiment *SIM* 6 (16) and appears comparable in size to similar estimates for other countries (17).

In addition to the decrease in income relative to history, there would have been larger negative (or smaller positive) current account balances. Thus, the fixed nominal exchange rate policy, although attainable and capable of achieving continued success in terms of decreasing the rate of inflation, would probably have not been desirable given its costs in terms of GDP and its effects on the trade balance (the actual cumulated balance between 1979 and 1986 is $-25,500$ billion while in the simulation it is equal to $-35,425$).

4.2.2 The Role of the Interest Rate

The interest rate, as an instrument of disinflation, is decidedly less effective and costlier than the exchange rate. The basic reason lies in the transmission mechanisms of interest rate changes to prices, mechanisms that operate through aggregate demand and unemployment. As these have, in turn, a relatively small and delayed impact on prices and wages, it follows that changes in the interest rate mainly affect the level of income and only slowly produce effects on inflation.

In order to evaluate the effects produced by the rule, it has been contrasted with two alternatives, one that simulates no active disinflationary policy (*SIM* 7), and one that simulates a very rapid, rather than gradual, adjustment of the interest rate (*SIM* 8).

The alternative in *SIM* 7 consists of maintaining the real interest rate (the nominal rate on short-term Treasury bills, deflated with the expected inflation rate of consumer prices) constant at 2.5%, the level reached in 1981.1, and equal to that experienced on average between the first and the second oil shock. Compared with history, this amounts to an expansionary policy, as the real rate was, on average,

---

(16) The sacrifice ratio is computed as $\Sigma (y'-y) / \Sigma (q'-q) / N$ where y is the log of GDP, q the inflation rate and N the number of quarters in the simulations. The "prime" indicates simulated values. No discounting is applied.

(17) On this issue, see GORDON-KING [6].

approximately 6.5% between 1981 and 1986. The inflation rate would have been slightly higher, with the larger difference relative to the historical value, 1.8 points, occurring in 1986 (Table 5). A gain in terms of GDP, approximately 0.7% on an annual basis, would have also occurred. But at the same time we would have experienced balance of payments deficits that would have made the policy unsustainable. In fact, in particular after the end of 1983, the ratio of reserves to imports worsens dramatically until it reaches 13% at the end of 1986, just above the historical minimum. This result is very important as it places the interest rate rule in the proper perspective, as indispensible to enforce the exchange rate rule. In other words, the exchange rate rule and the interest rate rule constitute a whole package that characterizes economic policy in the period.

It is also interesting to analyze the effects of a "shock treatment" policy, simulating an increase of the nominal interest rate that implies a much faster rise of the real rate compared to the increase that occurred between early 1980 and the end of 1982. In experiment *SIM* 8, the interest rate rule has been changed to attain in 1980.1 the level of the real rate actually reached in 1981.4. After this date, the historical pattern was maintained. This is a very violent shock as it requires an increase of almost 10 points in the nominal rate in the first quarter of 1980. The effect of the experiment is vividly depicted in Graph 5, which shows the differential between simulated and actual inflation rates and the logarithm of the ratio of simulated to actual GDP. The effect on inflation is rather small and temporary. The effect on the level of income is, instead, large and amounts to a cumulated loss of approximately 9%. The sacrifice ratio computed for such a policy is very high, equal to 25. This confirms what was noted, at the beginning of this section, about the limited effectiveness of the interest rate policy in controlling inflation.

4.2.3 Precommitment and the Disinflation of 1984

Income policies have been advocated many times and from several parts, as a means for reducing inflation more quickly and incurring in lower costs. Nevertheless, they were not implemented until the agreement of January 1983 (*Protocollo d'Intesa*) and the

GRAPH 5

INFLATION/GDP TRADE-OFF (*)

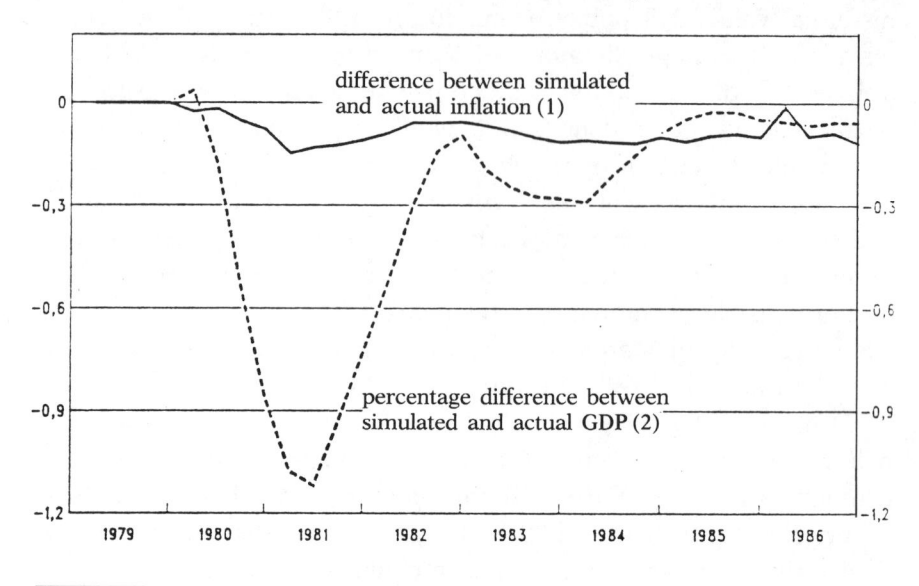

[Graph showing difference between simulated and actual inflation (1) and percentage difference between simulated and actual GDP (2), from 1979 to 1986, y-axis from -1.2 to 0]

(*) Simulation with instantaneous adjustment of the real interest rate on short-term Treasury Bills in 1980.1 at its actual level in 1981.4;
(1) Percentage change, consumer price index, seasonally adjusted;
(2) level, seasonally adjusted.

more comprehensive precommmitment experiment in 1984. Then, for the first time, aggregate demand and exchange rate management policies were supported by policies aimed at controlling nominal wages and influencing inflationary expectations, so as to attain a faster adjustment of prices and wages to changes in nominal demand and to avoid income losses. To achieve this goal, a target inflation rate was announced, in order to affect expectations. Accordingly, as discussed in Section 2, the pattern of nominal wages was set in advance as a function of the target inflation rate.

The results of the precommitment policy can be evaluated from the experiment *SIM* 9; the evolution of the actual rate of inflation versus the rate simulated in absence of the precommitment agreement is reported in Table 5. The individual contribution of this policy appears to have been rather small in 1984. In fact, the effect on inflation amounts to half a percentage point in 1984, when the largest decline in the annual rate of inflation was achieved, and the lagged effect in 1985 is a little less than one point (18). This result follows in part from the specification of the Phillips curve in the model; in fact, the presence of a catch-up term in the equation implies that expectations affect wage growth only transitorily.

4.2.4 The Deceleration of the Disinflation Process in 1985 and the Decline of Inflation in 1986

With the interruption of the anti-inflationary policy that had been followed in 1984, the process of disinflation comes to a pause in 1985. As shown in Table 7, there is a recrudescence of the endogenous component of inflation. After reaching a minimum of 2.2% in the first quarter of 1985, endogenous inflation rises steadily until the third quarter of 1986, when it settles at 7%, above the actual level of overall inflation. Only the strong deflationary impulse which originated abroad in 1986 allowed the overall inflation rate to decline further.

This new increase of internal inflation is probably caused by the effect of large changes in firms' profit margins, changes that cannot be explained, given the model estimates, by cyclical fluctuations in economic activity, or the evolution of competitiveness. These changes are captured by large forecast errors in 1985 and 1986 in the equations of value-added deflators. The rise in profit margins appears to be relevant particularly for the service sector, whose deflator (Table 8) rose in 1985 and 1986 at rates markedly higher than the cost components affecting it.

This interpretation, at the current stage of our analysis, is to be considered preliminary as it was reached by exclusion. It would require a specific test. In particular, the reliability of the data for the value-added deflators in 1986 should be further investigated, given the most recent revisions in the National Income Accounts. If the data are

---

(18) GUISO-MAGNANI [9] attain results decidedly more favorable to precommitment. Using an equation for inflation where the expected rate appears directly with a coefficient equal to 0.4, they estimate a decline of inflation in 1984 due to the announcement effect alone equal to 0.6 percentage points.

TABLE 8

VALUE-ADDED DEFLATOR OF THE SERVICE SECTOR
AND ITS MAIN COMPONENTS
(average annual percentage change)

Year	Value-added deflator of the service sector (1)(2)	Unit labor cost in the service sector (1)	Output price in the manufacturing sector
1980	22.3	19.2	23.2
1981	19.6	16.3	17.8
1982	16.9	16.4	17.7
1983	11.4	11.7	15.6
1984	9.1	9.8	9.6
1985	11.1	7.1	7.4
1986	9.2	3.7	7.4

(1) Value-added defined as excluding indirect taxes and production contributions.
(2) Service sector defined as excluding the production of services whose prices are publicly regulated.
Source: BANCA D'ITALIA, DATA BANK, *Quarterly Econometric Model* (as of April 1987).

indeed not subject to error, then the causes of this increase in profit margins should be highlighted, examining in particular the role played by the sudden and substantial drop in raw material prices and by domestic factors.

5. - Conclusions

The following conclusions on the 1980-1986 disinflation experience can be drawn:

a) the evolution of foreign prices had a considerable effect on the rate of inflation in Italy, owing to the degree of openness of the economy and the existence of widespread indexation mechanisms;

b) the process of disinflation was, however, made possible by the continuous decline of the internal component rather than by diminishing foreign impulses, with the exception of 1986 when the effects of the large drop in oil prices and of the US dollar depreciation were very significant;

c) monetary policy, including both exchange rate and interest rate management, played a decisive role in curbing inflation. In particular, the adoption of an accommodating exchange rate behavior would have implied a much higher inflation path, relative to history, with small gains in terms of income. On the other hand, maintaining the exchange rate parity that existed prior to the second oil shock would have led to large income losses and foreign account imbalance, which probably would have made such a policy unsustainable;

d) the policy of "precommitment" to target inflation rates announced by the government appears to have had only limited effects in curbing inflation. This policy had the likely effect, however, of preventing the rise of autonomous impulses to the inflation process, which would have increased the costs of the monetary and exchange rate policies and hampered their implementation.

APPENDIX

The Endogenization of International Variables

Some of the simulations reported in this paper were conducted under specific assumptions about international prices and exchange rates. In particular, the following alternatives were considered:

1) absence of the second oil shock of 1979 and the negative shock of 1986. In this case, the deflator of Italian imports of energy products denominated in dollars (*PIMPENW*) was assumed constant at its 1979.1 level.

2) absence of the appreciation and consequent depreciation of the dollar in the 1980s. In this case we assumed invariance of the dollar/deutsche mark rate (*EXDMUS*) at its 1979.1 level, keeping all other bilateral rates of the deutsche mark at their actual values except for the lira, whose rate varies with the reaction function.

3) all international prices constant at their 1979.1 levels.

It was chosen to allow the two exogenous variables *PIMPENW* and *EXDMUS*, which we assumed mutually independent for the sake of simplicity, to affect the other international prices. Unconstrained equations were then estimated for the three fundamental price variables — i.e., the deflator of Italy's imports of agricultural (*PIMPAGW*) and nonagricultural products (*PIMPMPW*), both denominated in dollars, and the export price of manufactured goods of Italy's trade partners (weighted with export shares and expressed in national currencies, *PALTRI*) — with the form:

$$(1) \quad P = \alpha + \sum_{i=1}^{2} \beta_i P_{-i} + \sum_{i=0}^{2} \gamma_i PE_{-i} + \sum_{i=0}^{2} \delta_i EX_{-i} + \sum_{i=0}^{2} \epsilon_i D_{-i},$$

where $P = PIMPMPW$, $PIMPAGW$, $PALTRI$, $PE = PIMPENW$ (import price of energy products, in dollars), $EX = EXDMUS$ (deutsche mark/dollar exchange rate), and $D = DOMW$ (world demand, i.e., total world imports excluding Italy, in constant dollars).

All variables are expressed as percentage changes and the estimation period is 1976.1-1985.4. Notice that equation *(1)* attempts to capture the effects stemming from changes in the price of energy products and in the dollar/deutsche mark rate after allowing for the effects of world demand and the endogenous autoregressive structure. Thus, essentially reduced forms have been estimated.

The estimates of equation *(1)* are reported in Table A1. From the unrestricted estimation restricted estimates were obtained, eliminating the variables with insignificant coefficients on the basis of the usual statistical tests. In addition to testing these restrictions, the forecasting stability of the estimates was also verified by comparing them with estimates obtained on a sample ending in 1982.4. In both cases, satisfactory results were obtained.

From the restricted estimates there does not appear to be any endogenous dynamics for agricultural and nonagricultural raw materials. The elasticities of these variables with respect to energy prices and the dollar/deutsche mark rate are similar, equal to 0.3/0.4 and $-0.4/-0.5$, respectively. On the contrary, the elasticity with respect to world demand is much higher for nonagricultural raw materials (approximately unity versus 0.4 for agricultural raw materials).

On the other hand, changes in the dollar/deutsche mark rate produced only a small and transitory effect on the export prices of manufactured goods of Italy's major trade partners. The elasticities with respect to energy prices are small (0.05 for the impact elasticity and 0.08 in the long run), as are the elasticities with respect to world demand (0.09 and 0.14).

To take into account the fact that world demand itself may be affected by changes in the relative price of energy (defined with respect to the price of manufactured goods of Italy's trade partners), for the period 1976.1-1985.4 an equation has been estimated of the form (See Table *A2*):

$$(2) \qquad D = \alpha + \sum_{i-1}^{4} \beta_i D_{-i} + \gamma t + \sum_{i=0}^{4} \delta_i (PE/P)_{-i},$$

where $D = DOMW$, $t = TREND$ (linear trend, equal to 1 in 1960.1), $P = PALTRIW$ (export price of manufactured goods of Italy's major competitors, denominated in dollars, weighted with export shares), and $PE = PIMPENW$.

TABLE A1

INTERNATIONAL PRICES (*)

	PIMPMPW NR	PIMPMPW R	PIMPAGW NR	PIMPAGW R	PALTRI NR	PALTRI R
c_0	−0.006 (0.493)	−0.006 (0.599)	−0.003 (0.378)	0.002 (0.467)	0.006 0.701)	(2.962)
c_1	0.325 (1.320)	0.402 (2.660)	0.287 (2.520)	0.305 (3.686)	0.039 (1.619)	
c_2	−0.002 (0.008)		−0.057 (0.460)		0.030 (1.293)	0.051 (2.528)
c_3	0.153 (0.675)		−0.067 (0.525)		−0.019 (0.815)	
c_4	−0.540 (2.260)	−0.539 (3.111)	−0.492 (3.939)	−0.469 (4.942)	0.125 (5.126)	0.105^a (4.924)
c_5	0.023 (.071)		0.236 (1.274)		−0.050 (1.443)	0.105^a (4.924)
c_6	−0.211 (0.765)		−0.151 (0.908)		−0.012 (0.414)	
c_7	0.374 (.678)		0.120 (0.418)		0.066 (1.019)	
c_8	0.921 (1.891)	0.950 (2.208)	0.291 (1.834)	0.396 (1.680)	0.104 (1.936)	0.086 (1.648)
c_9	−0.314 (0.552)		0.156 (0.528)		−0.011 (0.170)	
c_{10}	0.071 (0.352)		0.161 (0.833)		0.429 (2.164)	0.375 (2.610)
c_{11}	−0.074 (0.354)		−0.024 (0.135)		0.181 (0.960)	
R^2	0.453	0.415	0.639	0.574	0.687	0.582
SER	0.057	0.052	0.030	0.028	0.006	0.006
DW	1.923	1.965	2.003	1.694	1.936	1.932
MLM1	0.001	0.003	0.619	0.560	0.000	0.010
MLM4	0.000	0.183	0.583	0.527	0.093	0.003
ARCH1	0.400	0.007	0.025	1.229	1.028	4.478
N	40	40	40	40	40	40
CHOW	0.359	0.373	0.834	0.782	0.803	1.121
F		0.240		0.632		1.345

(*) Estimated equation:

$\Delta \log(P) = c_0 + c_1 \Delta \log(PIMPENW) + c_2 \Delta \log(PIMPENW_{-1})$
$\quad + c_3 \Delta \log(PIMPENW_{-2})$
$\quad + c_4 \Delta \log(EXDMUS) + c_5 \Delta \log(EXDMUS_{-1}) + c_6 \Delta \log(EXDMUS_{-2})$
$\quad + c_7 \Delta \log(DOMW) + c_8 \Delta \log(DOMW_{-1}) + c_9 \log(DOMW_{-2})$
$\quad + c_{10} \Delta \log(P_{-1}) + c_{11} \Delta \log(P_{-2})$

where $P = PIMPMPW$, $PIMPAGW$, and $PALTRI$, respectively.
Estimation period: 1976.1-1985.4.

Variables: See text.

Legend: (see Table 1): NR = unrestricted model; R = restricted model; F = test of the restrictions, distributed as F with R, $N-K$ degrees of freedom, where R is the number of restrictions;
[a] Coefficients restricted to be of opposite sign and equal value.

TABLE A2

WORLD DEMAND (*)

	NR	R
c_0	2.318	2.332
	(3.893)	(4.941)
c_1	0.004	0.004
	(3.717)	(4.807)
c_2	0.008	
	(0.165)	
c_3	0.011	
	(0.137)	
c_4	−0.018	
	(0.228)	
c_5	−0.077	−0.075
	(0.990)	(3.516)
c_6	0.003	
	(0.058)	
c_7	0.613	0.591
	(3.326)	(4.043)
c_8	0.330	0.344
	(1.659)	(1.978)
c_9	−0.363	−0.313
	(1.764)	(2.601)
c_{10}	−0.039	
	(0.254)	
R^2	0.981	0.980
SER	0.015	0.014
DW	2.025	1.943
MLM1	0.841	0.028
MLM4	9.916	8.955
ARCH1	0.177	0.378
N	40	40
CHOW	0.666	0.572
F		.074

(*) Estimated equation:

$\log(DOMW) = c_0 + c_1\ TREND + c_2 \log(PIMPENW/PALTRIW)$
$\quad + c_3 \log(PIMPENW/PALTRIW)_{-1} + c_4 \log(PIMPENW/PALTRIW)_{-2}$
$\quad + c_5 \log(PIMPENW/PALTRIW)_{-3} + c_6 \log(PIMPENW/PALTRIW)_{-4}$
$\quad + c_7 \log(DOMW_{-1}) + c_8 \log(DOMW_{-2}) + c_9 \log(DOMW_{-1})$
$\quad + c_{10} \log(DOMW_{-4})$.

Estimation period: 1976.1-1985.4.

Variables: see test.

Legend: see Tables 1 and A1.

The variables are expressed in logs with the exception of the trend. In the restricted equation used in the simulations, there is a significant effect of the relative price of energy, lagged three quarters. The impact elasticity with respect to the latter variable is equal to 0.075 and the long run elasticity is equal to 0.2. The estimates appear stable, however, one can notice fourth-order autocorrelation that might indicate problems in seasonal adjustment (correcting for autocorrelation leaves the estimates basically unchanged).

The value of the elasticity of world demand to the relative price of energy seems rather high; since the ratio PE/P rose by 100% between 1979.1 and 1981.3, the estimates imply that the resulting decline in world demand was equal to 20%. According to estimates obtained at the OECD (Larsen and Llewellyn [10]), the elasticities of OECD imports with respect to oil prices are in the range of 0.03-0.09 in the first year and 0.10-0.22 after 3 years, when simulating a 25% drop in oil prices and alternative economic policy reactions.

In order to maintain consistency between the international variables used in the model, the remaining international prices and real variables were also endogenized. The following international prices have been endogenized as historical ratios:

1) the price of gasoline and other oil products in the EEC, as functions of the import price of energy products.

2) agricultural prices in the EEC, as a function of the import price of agricultural products.

3) consumer prices in the Mediterranean area and in the OECD countries, and the price of manufactured goods of Italy's trade partners, weighted with import shares, as functions of the export price of manufactured goods of Italy's trade partners, weighted with export shares (*PALTRI*).

The interest rate on the eurodollar and on the other eurocurrencies was also endogenized, by adding the difference between simulated and actual inflation rates of the export price of manufactured goods of Italy's trade partners to their historical values.

Finally, GDP of the OECD countries has been endogenized as a function of world demand.

BIBLIOGRAPHY

[1] BANCA D'ITALIA: «Modello trimestrale dell'economia italiana, vol. I e vol. II», Banca d'Italia, *Temi di discussione*, n. 80, 1986.
[2] BASEVI C. - CALZOLARI M. - COLOMBO C.: «Monetary Authorities Reaction Functions and the European Monetary System», in D.R. Hodgman (ed.): *The Political Economy of Monetary Policy: National and International Aspects*, Federal. Reserve Bank of Boston, Conference Series n. 26, 1983.
[3] BINI SMAGHI L. - VONA S.: «Le tensioni commerciali nello SME: il ruolo delle politiche di cambio e della convergenza economica», Banca d'Italia, *Contributi all'analisi economica*, n. 2, 1986.
[4] ENGLE R.F. - HENDRY D.F. - RICHARD J.F.: «Exogeneity», *Econometrica*, March 1983.
[5] GIANNINI C.: «Sul concetto di variabile esogena e sulla previsione condizionale», *Politica Economica*, August 1986.
[6] GORDON R.J. - KING S.R.: «The Output Cost of Disinflation in Traditional and Vector Autoregressive Models», *Brookings Papers of Economic Activity*, pp. 205-44, 1982.
[7] GRESSANI D. - GUISO L. - VISCO I.: «Il rientro dell'inflazione: un'analisi con il modello econometrico della Banca d'Italia», Banca d'Italia, *Contributi all'analisi economica*, n. 3, 1987.
[8] GUISO L.: «Prime rate, aspettative a tasso di sconto. Una verifica dell'ipotesi di razionalità», Banca d'Italia, *Contributi all'analisi economica*, n. 2, 1986.
[9] GUISO L. - MAGNANI M.: «1981-1984 - Perché l'inflazione è calata?», Banca d'Italia, 1985.
[10] LARSEN F. - LEWELLYN J.: «Simulated Macroeconomic Effects of a Large Fall in Oil Prices», Oecd, Economics and Statistics Department, Working paper n. 8, 1983.
[11] LUCAS R.F.: «Econometric Policy Evaluation: A Critique», in K. Brunner-A.H. Meltzen (eds): *The Phillips Curve and Labor Markets*, Carnegie-Rochester Conference Series, vol. 1, Amsterdam, North-Holland, 1976.
[12] RUBINO P. - VISCO I.: «Politica tariffaria e prezzi al consumo: un'analisi applicata», Cnr, Progetto finalizzato *Struttura ed evoluzione dell'economia italiana*, Sottoprogetto n. 3, Tema 3, Linea E, 1987.
[13] SIMS C.A.: «Policy Analysis with Econometric Models», *Brookings Papers on Economic Activity*, n. 1, 1982.
[14] VISCO I.: *Price Expectations in Rising Inflation*, Amsterdam, North-Holland, 1984.
[15] VISCO I.: «Inflation Expectations: The Use of Italian Survey Data in the Analysis of Their Formation and Effects on Wage Changes», paper presented at *Workshop on Price Dynamics and Economic Policy*, OECD, September 1984.
[16] VISCO I.: The Indexation of Earnings in Italy: Sectoral Analysis and Estimates for 1978-1979, Selected Papers *Rivista di politica economica*, n. 13.

The Government Budget and the Italian Economy During the 1970s and 1980s: Causes of the Debt, Strategy for Recovery, and Prospects for Restructuring (*)

Mario Baldassarri
Università «La Sapienza», Roma

M. Gabriella Briotti
Centro Studi Confindustria, Roma

Introduction

Our purpose is to analyze the structure and the workings of the Italian government budget over the last 20 years and to focus on the contradictory and turbulent relationship which existed between the government budget and Italy's economic performance during the same period.

This paper is divided into four sections.

In the first section, we compare the structure and the workings of the Italian government budget with those of the major industrialized countries. In the second section, we attempt to measure the economic impact of the government budget in the short term and its medium and long-term structural effects on resource allocation and on levels of productive capacity and full employment in this country.

As imbalances in government budget must be offset by financial backing at year end and over the long term, in the third section we seek to attach a name and a face to the lender of last resort.

In the fourth section, we explore the causes of Italy's public debt in order to identify the crucial strategy for recovery. However, we surmise that it is absolutely essential to end the dangerous practice of permitting unrestrained increases in public expenditure, which are

---

(*) This article previously appeared in *Rivista di politica economica*, Roma, Sipi, July-August 1990.

either openly supported or tacitly accepted, to precede budget allocations, resulting in complex tax hikes that draw cries of pain from several quarters. Since these increases are always decided upon under emergency conditions and do little to raise government revenue by broadening the tax base, they continue to worsen the tax burden placed on the industrial system (firms and workers), thus undermining the development potential of the system. Against this background, expenditure, deficit and debt continue unchecked, and the uneven distribution of the heavy tax load, is rife with inequality.

Synthesis: Some Conclusions

Comparison of Italy's budget policy in the 1970s with the parallel experiences of the other industrialized countries reveals how much earlier Italy had begun to implement a policy of increased government expenditure, with respect to the degree of development it had achieved. There was, however, a delay, relative to the other countries, in raising tax revenue to cover the rise in expenditure, per capita income being equal. Italy therefore seems to have increased expenditure on social services while refusing to promptly pay the price in terms of higher tax revenue.

The subsequent staggering rise in fiscal pressures during the 1980s, equal to some 10% of gross domestic product, essentially financed the servicing of a massive public debt rather than restoring equilibrium to flows of revenue and expenditure.

Our findings indicate that maintaining high budget deficits had substantial inflationary effects in the short term. Furthermore, budget deficit policies produced distortionary effects on resource allocation in the long term, thus reducing the potential for production and employment within the system.

It is common knowledge that inherent in the financing of the Italian public debt is a strong effect of income redistribution, both among successive generations and within the same generation, necessary in order to maintain financial equilibrium in the short and long term. Our findings point to holders of government securities as the providers of financial backing who until 1983 contributed to balance accounting budgets at year end by incurring heavy losses in real terms. Since then, the government budget has maintained financial equilibrium in accounting terms by transferring to future generations the cost of restoring true equilibrium to public sector accounts.

Our results also reveal the lack of coordination of revenue inflows and expenditure outflows from the early 1970s to the beginning of the 1980s as one of the major factors contributing to the high public debt. Bringing expenditure in line with available funds would have generated sufficient budget resources for a large expenditure policy.

The objectives pursued during the 1970s were predominantly oriented toward social services, while support for industrial activity played a minor role. As a consequence, our paper proposes a twofold strategy for recovery: as regards the profit-and-loss statement, a readjustment of expenditure to revenue by curtailing expenditure outflows; as regards the balance sheet, reducing liabilities (public debt) by reducing government assets through the sale of state property. In conclusion, it should be pointed out that the sale of a portion of assets in order to offset the current account deficit is an incorrect accounting operation. Its positive effects on the current deficit can be brought about by reduced interest payments on a smaller public debt.

1. - Structure and Workings of the Government Budgets in Italy and in the Other Major Industrialized Countries

The size, structure and workings of the governement budget cannot be correctly evaluated by examining the case of only one nation; rather, these characteristics must be confirmed by similar experiences in other countries of comparable size and economic outlook. Italy now occupies a permanent place within the leading group of the largest industrial economies, and it is therefore enlightening to compare this country's experience with the events of the five major industrialized countries, namely, United Kingdom, France, Germany, United States and Japan.

This section aims at providing a historical framework for government behaviour from 1970 to the present, beginning with the large aggregates of expenditure and revenue of general governments and their budget balances.

However, international comparison must take into account the degree of development achieved by each country: for this reason, we compared budget data not only for one year, but also over several years in which equivalent levels of per capita income were reached.

This comparative study then examines more detailed information regarding the major line items of the government budget and the breakdown by institutional sectors of expenditure and revenue items, of savings and of financial balances. Finally, this analysis provides information on the absorption of total domestic credit by the public sector.

1.1 *Expenditure, Revenue and Balances of General Government*

The government budget has grown significantly over the last twenty years in all industrialized countries. In the mid-1960s, total general government expenditure, accounted for 36% of GDP in European countries, and 20% and 25% in Japan and the United States respectively. During the 1980s these percentages climbed to approximately 50% in Europe and neared 35% in the United States and Japan. In the same period total revenues accounted for an increasing portion of GDP. (Table 1*A*).

More specifically, in the mid-1960s, public sector revenue and expenditure expressed as a ratio to GDP were both lower in Italy than in other European countries. The steadily increasing trend which followed drove total expenditure upward, by approximately 20% of GDP throughout recent years and resulted in a strong but limited rise in revenue of approximately 12% if GDP. In the late 1980s, then, the size of the Italian government budget seemed to be in line with those of other countries as far as the expenditure-GDP ratio was concerned, while the revenue-GDP ratio was apparently lower.

A look at the budget balances, general government current account savings and financial balance for all line items reveals an overall trend in all countries, from the early 1970s onward, toward lower general government savings and a worsening of the total financial balance (Table 1*B*).

TABLE 1A

REVENUE AND EXPENDITURE OF GENERAL GOVERNMENT
IN % OF GROSS DOMESTIC PRODUCT

Country	1965	1969	1974	1978	1980	1981	1982	1983	1984	1985	1986	1987	1988
Total revenue													
Italy............	27.2	27.9	27.8	30.2	33.4	34.5	37.1	38.9	38.0	38.3	39.3	39.6	40.2
United Kingdom	33.4	39.8	40.1	38.1	40.3	42.7	43.3	42.6	42.6	42.6	41.8	40.9	n.a.
France	38.4	39.8	39.4	42.3	47.0	47.7	48.5	49.2	50.2	50.4	48.1	48.6	48.2
Germany	35.3	38.3	41.5	43.3	45.2	45.3	45.8	45.4	45.6	45.9	45.2	44.7	44.1
United States	27.0	31.2	31.9	32.1	31.1	31.9	31.4	31.0	30.9	31.5	31.5	32.2	31.8
Japan	19.5	19.5	24.4	24.6	27.1	28.6	29.2	29.6	30.4	31.1	31.4	33.5	34.5
Total expenditure													
Italy............	31.7	31.7	35.1	42.6	42.2	46.1	48.4	49.8	49.8	51.1	51.2	51.0	51.1
United Kingdom	36.3	41.5	45.2	43.7	45.2	48.0	47.4	47.2	47.6	46.5	45.7	43.5	n.a.
France	38.2	39.6	39.6	45.1	48.6	51.3	53.0	54.0	54.7	55.0	52.6	52.3	51.3
Germany	36.1	37.4	43.1	46.4	48.8	49.7	49.7	48.7	48.3	47.8	47.2	47.1	46.9
United States	27.7	30.8	33.2	32.8	34.1	34.4	36.8	37.1	36.0	36.9	37.2	37.0	36.5
Japan	18.7	17.9	23.6	29.8	32.1	33.1	33.4	33.9	33.2	32.6	33.0	33.4	33.0

Sources: OECD: *National Accounts* 1981, 1990 and our estimates.

TABLE 1B

SAVINGS AND FINANCIAL BALANCES OF GENERAL GOVERNMENT IN % OF GROSS DOMESTIC PRODUCT

Country	1965	1969	1974	1978	1980	1981	1982	1983	1984	1985	1986	1987	1988
General government savings
Italy	0.0	0.7	−4.2	−6.7	−4.8	−7.2	−8.7	−7.0	−7.4	−7.2	−7.0	−6.6	−6.2
United Kingdom	2.4	6.8	1.0	−1.7	−1.6	−1.5	−1.4	−1.6	−2.1	−1.2	−1.2	−0.5	n.a.
France	4.7	4.6	3.6	0.3	2.2	0.1	−0.7	−1.4	−1.2	−1.3	−1.1	−0.6	0.2
Germany	4.9	6.1	4.0	2.0	1.8	0.5	0.4	0.7	1.4	2.0	1.8	1.2	0.7
United States	1.6	2.5	0.7	0.8	−1.6	−1.3	−4.2	−5.0	−3.9	−4.0	−4.3	−3.3	−3.3
Japan	5.4	5.9	6.3	1.4	2.6	3.1	2.8	2.3	3.3	4.3	4.2	6.0	7.5
General government financial balances
Italy	−4.5	−3.4	−7.2	−9.9	−8.6	−11.4	−13.6	−10.6	−11.6	−12.5	−11.7	−11.2	−10.6
United Kingdom	−2.0	−0.6	−3.9	−4.3	−3.5	−3.9	−2.8	−3.4	−3.8	−2.7	−2.8	−1.4	n.a.
France	0.7	0.9	0.6	−1.8	0.0	−1.9	−2.8	−3.2	−2.8	−2.8	−2.7	−2.0	−1.4
Germany	−0.6	1.1	−1.4	−2.7	−2.9	−3.7	−3.3	−2.5	−1.9	−1.1	−1.3	−1.8	−2.1
United States	0.3	1.6	0.4	0.6	−1.5	−1.1	−4.0	−4.9	−3.8	−4.2	−4.4	−3.5	−3.6
Japan	n.a.	n.a.	0.4	−5.9	−4.4	−3.8	−3.6	−3.7	−2.1	−0.8	−1.0	0.7	2.1

Sources: OECD: *National Accounts* 1981, 1990 and our estimates.

Within this broader context, two situations are readily apparent. On the one hand, Japan, Germany and, to some extent, France witnessed a lowering of their general government savings, which nonetheless remained positive, and a worsening of the total balance linked mainly to capital account expenditure patterns; on the other hand, the general government savings of Italy, the United Kingdom and, more recently, the United States remained consistently negative (as the government budget depleted national savings). For these countries, the worsening of the total financial balance was largely due to the growth of current account expenditure. Italy differed from the second group of countries in several ways. First, negative values for public sector savings in Italy were manifest from as early as 1971 onward; slowly but steadily, they depleted national savings, maintaining levels of -7% / -8% of GDP during much of the 1980s. Second, the worsening of the financial balance appeared to be primarily due to current account expenditure throughout the seventies; in the 1980s, capital account expenditure began to play a more significant role. However, we should remember that at that time, capital account expenditure in the Italian government budget consisted primarily of transfers to cover budget year losses of public sector firms and agencies, hardly a true example of capital accumulation for the public sector or industry.

Yet another Italian feature concerned the major sources of revenue, direct taxes, indirect taxes and social security contributions (Tables 2*A* and 2*B*). With the basic and constant fact that social security contributions were by far higher than those of the other countries, went an increasing disproportion between direct and indirect taxes, both of which were increased later than in the other countries, especially in Europe. At the beginning of the 1970s, in fact, the portion of revenue accounted for by direct taxes in Italy seemed smaller than those of the United Kingdom and Germany in particular; the opposite is true with respect to indirect taxes. Italy was therefore pursuing a strategy of correcting the disequilibrium by increasing direct taxes more than indirect taxes. Espressed as a ratio to GDP, toward the end of the decade, the former exceeded the latter.

This phenomenon continued in the 1980s, while other countries changed course toward higher indirect taxes. To be sure, these

TABLE 2A

BREAKDOWN OF GENERAL GOVERNMENT EXPENDITURE

Country	1965	1969	1974	1978	1980	1981	1982	1983	1984	1985	1986	1987	1988
Capital account expenditure (in % of total expenditure)													
Italy	11.1	11.1	9.5	9.6	10.2	10.4	10.6	10.2	10.0	11.6	10.1	10.0	9.8
United Kingdom	14.9	20.5	13.7	8.9	7.7	8.5	6.3	6.8	6.7	6.3	6.6	5.2	n.a.
France	11.8	11.4	9.3	9.3	9.5	8.9	8.7	8.1	7.8	7.9	8.3	8.0	8.7
Germany	15.8	13.9	13.0	11.0	12.0	10.6	9.6	8.9	9.0	8.7	8.6	8.4	8.1
United States	8.3	6.8	6.0	4.9	5.0	4.6	4.2	4.0	3.9	4.5	4.1	4.5	4.7
Japan	24.1	24.0	23.3	21.8	22.2	21.4	20.3	19.0	18.0	17.4	17.3	18.1	18.8
Current account expenditure (in % of total expenditure)													
Italy	88.9	88.9	90.5	90.4	89.8	89.6	89.4	89.8	90.0	88.4	89.9	90.0	90.2
United Kingdom	85.1	79.5	86.3	91.1	92.3	91.5	93.7	93.2	93.3	93.7	93.4	94.8	n.a.
France	88.2	88.6	90.7	93.1	90.5	91.1	91.3	91.9	92.2	92.1	91.7	92.0	91.3
Germany	84.2	86.1	87.0	89.0	88.0	89.4	90.4	91.1	91.0	91.3	91.4	91.6	91.9
United States	91.7	93.2	94.0	95.1	95.0	95.4	95.8	96.0	96.1	95.5	95.9	95.5	95.3
Japan	75.9	76.0	76.7	78.2	77.8	78.6	79.7	81.0	82.0	82.6	82.7	81.9	81.2

Source: OECD: *National Accounts* 1981, 1990 and our estimates.

TABLE 2B

BREAKDOWN OF GENERAL GOVERNMENT REVENUE

Country	1965	1969	1974	1978	1980	1981	1982	1983	1984	1985	1986	1987	1988
Direct taxes (in % of total revenue)													
Italy	21.0	21.6	21.3	28.7	28.8	31.8	32.2	31.8	33.1	33.9	32.8	33.7	33.4
United Kingdom	35.4	37.2	40.5	37.6	34.8	34.9	35.3	35.5	35.7	36.1	35.1	34.9	n.a.
France	16.5	17.3	18.8	18.5	17.9	18.1	18.2	18.2	18.6	18.4	19.3	19.4	18.9
Germany	28.1	28.0	31.3	30.1	28.1	26.9	26.4	26.3	26.6	27.3	27.2	27.4	27.5
United States	47.6	49.0	44.7	45.7	46.0	44.5	42.7	41.6	41.2	41.4	41.4	43.5	41.9
Japan	38.5	39.5	45.0	37.4	39.8	39.4	39.0	39.3	39.0	39.2	39.2	38.8	38.4
Indirect taxes (in % of total revenue)													
Italy	38.8	38.2	33.1	23.8	25.9	24.0	23.3	23.5	24.4	23.4	23.2	24.0	25.1
United Kingdom	40.7	40.8	33.6	35.9	38.6	38.7	38.1	37.5	37.3	36.7	38.7	39.1	n.a.
France	43.9	40.9	36.5	33.1	31.1	30.3	30.2	29.8	29.9	29.6	30.3	30.4	30.3
Germany	38.3	37.0	29.8	29.4	28.9	28.4	27.6	28.2	28.2	27.4	27.1	27.4	27.6
United States	33.7	29.5	28.5	26.0	25.5	26.2	26.4	27.2	27.3	26.7	26.4	25.4	25.7
Japan	37.5	35.1	28.3	27.9	27.2	26.4	25.8	24.9	25.3	25.3	24.3	24.7	24.5
Social security contributions (in % of total revenue)													
Italy	35.6	36.3	41.1	38.0	34.4	33.2	33.6	32.1	31.5	31.2	31.6	31.4	30.9
United Kingdom	14.2	12.1	15.0	16.1	15.0	14.7	15.1	16.2	16.2	16.0	16.4	16.7	n.a.
France	30.7	32.4	35.6	39.3	37.9	37.2	37.7	38.4	38.2	38.2	39.3	39.6	39.7
Germany	26.4	27.2	31.0	32.4	34.4	35.6	32.4	35.4	35.3	35.4	35.9	36.3	36.7
United States	16.1	18.7	23.0	23.6	19.9	20.4	21.3	21.8	22.4	22.6	22.9	22.3	23.6
Japan	19.7	20.8	21.1	27.8	26.9	27.2	27.5	27.6	26.8	26.6	26.8	25.8	25.9

Sources: OECD: *National Accounts* 1981, 1990 and our estimates.

comparisons refer to the tax revenues levied in accordance with legal and institutional procedure which, if complied with, would produce entirely different results. However, tax evasion and tax avoidance seriously limit this type of country comparison, and place particular strain on the burden of revenue increases borne by those compelled to form the tax base. The situation created is extremely complex and contradictory, since the necessity for higher taxes may add more inequality to the system. Further, the fact that "wage and salary taxation" forms a relevant part of the tax collection mechanism shifts the burden onto the industrial system and complicates the income distribution problem, undermines competitiveness, inhibits the investment and development process, and actually reduces the possibility of raising tax revenues.

Some caution is needed to qualify the conclusions previously formed, on the basis of accounting data, regarding the levels of revenue and expenditure reached in the Italian government budget compared with those of other countries. In fact, despite the significant progress achieved, Italian per capita GDP has remained consistently low compared with the other countries (Table 3).

In 1974, Italian per capita income equalled the levels reached in France and Germany in 1971 and in the United Kingdom in 1972. The gap between Italy and the first two countries continued to widen, albeit at a moderate pace, throughout the 1980s, and in 1989 it narrowed once again to delays of two and approximately four years respectively behind France and Germany. Italy closed the space with the United Kingdom in 1980, when British per capita income dipped slightly below the Italian figure; it remained at approximately the same level until 1987, and then rose slightly during 1988-1989. From particularly low levels of income in 1970, Japan ended its long march in 1988, outdistancing all of the European countries, including Germany, and yet remaining some distance away from matching United States per capita income.

As we can see, then, it is useful to juxtapose the Italian budget policy with the parallel experiences of other countries at the same levels of development. This comparison is made in the tables which set out the total expenditure, total revenue, savings and financial balances recorded in Italy in 1974 (Table 4*A*), in 1981 (Table 4*B*), in 1985

TABLE 3

PER CAPITA GROSS DOMESTIC PRODUCT AT MARKET PRICES IN CONSTANT 1980 DOLLARS AND PURCHASING POWER

Year	Italy	United Kingdom	France	Germany	United States	Japan
1970	6,153.8	6,536.5	6,660.9	6,868.3	10,006.0	5,614.0
1971	6,213.9	6,672.4	6,914.4	6,991.9	10,201.0	5,797.4
1972	6,367.7	6,804.2	7,157.4	7,242.6	10,603.0	6,141.0
1973	6,767.8	7,338.8	7,486.4	7,544.7	11,005.0	6,533.9
1974	7,003.4	7,263.0	7,669.5	7,556.1	10,826.0	6,366.9
1975	6,710.7	7,215.3	7,613.1	7,462.5	10,611.0	6,452.6
1976	7,071.4	7,481.3	7,904.7	7,905.5	11,022.0	6,686.2
1977	7,202.5	7,562.9	8,123.1	8,155.6	11,397.0	6,971.8
1978	7,343.4	7,844.6	8,358.3	8,399.6	11,853.0	7,260.9
1979	7,681.4	8,011.0	8,592.9	8,743.8	11,954.0	7,574.2
1980	7,964.3	7,843.0	8,688.0	8,833.3	11,804.0	7,848.5
1981	8,042.2	7,744.9	8,740.8	8,831.3	11,946.0	8,092.5
1982	8,042.2	7,846.3	8,914.7	8,780.5	11,520.0	8,265.6
1983	8,100.3	8,142.8	8,935.2	8,943.6	11,855.0	8,469.0
1984	8,333.9	8,273.8	9,016.9	9,232.4	12,586.0	8,836.8
1985	8,552.7	8,550.6	9,129.6	9,438.9	12,937.0	9,197.7
1986	8,785.3	8,791.9	9,281.7	9,650.9	13,195.0	9,361.2
1987	9,040.5	9,097.5	9,444.4	9,827.2	13,535.0	9,715.2
1988	9,248.5	9,672.5	9,765.6	10,171.4	13,903.0	10,276.9
1989	9,511.6	9,914.7	10,051.5	10,460.2	14,179.0	10,737.3

Sources: OECD: *National Accounts* 1989; WEFA: *Historical data.* 1990.

(Table 4*C*) and in 1989 (Table 4*D*) and the situations of other countries at the same level of per capita GDP. The same criterion is used to evaluate the major line items of revenue and expenditure in Tables 5*A*, 5*B*, 5*C* and 5*D*.

The figures reveal that the Italian government deficit which could be attributed, using accounting data, to an undersized tax base are more correctly defined as a problem owing "in equal measure" to excessive government expenditure and to a shortage of revenue.

EXPENDITURE AND REVENUE OF GENERAL GOVERNMENT IN
SELECTED COUNTRIES, PER CAPITA GDP BEING EQUAL
(in % of GDP)

TABLE 4A

Country	Year	Total expenditure	Total revenue	General government savings	Financial balance
Italy.................	1974	35.1	27.8	−4.2	−7.2
United Kingdom	1972	38.9	35.9	2.5	−1.8
France	1971	38.3	38.3	3.9	0.7
Germany	1971	38.5	37.8	5.6	−0.2
Japan	1977	28.7	24.4	2.3	−3.8

Source: OECD: *National Accounts,* 1983 and 1989.

TABLE 4B

Country	Year	Total expenditure	Total revenue	General government savings	Financial balance
Italy.................	1981	46.1	34.5	−7.2	−11.4
United Kingdom	1979	42.9	38.4	−1.2	−3.2
France	1976-1977	44.6	43.8	3.5	−0.8
Germany	1976-1977	48.5	45.1	1.9	−2.9
Japan	1981	33.1	28.6	3.1	−3.8

Source: OECD: *National Accounts,* 1983 and 1989.

TABLE 4C

Country	Year	Total expenditure	Total revenue	General government savings	Financial balance
Italy.................	1985	51.1	38.3	−7.2	−12.5
United Kingdom	1985	46.7	42.7	−1.3	−2.9
France	1980	47.0	47.0	3.8	0.0
Germany	1978	46.4	43.3	2.0	−2.7
United States	1965	27.7	27.0	1.6	0.3
Japan	1983	33.9	29.6	2.3	−3.7

Source: OECD: *National Accounts,* 1983 and 1989.

TABLE 4D

Country	Year	Total expenditure	Total revenue	General government savings	Financial balance
Italy	1989	52.3	42.1	−5.7	−10.2
United Kingdom	1988	40.5	39.5	1.3	−1.3
France	1987	52.9	50.5	1.0	−2.4
Germany	1985-1986	47.5	45.5	1.9	−1.2
United States	1965	27.7	27.0	1.6	0.3
Japan	1986-1987	33.2	32.3	5.1	−0.2

Sources: OECD: *National Accounts*, 1983 and 1989; BANK OF ITALY: *Annual Report*, 1990; our estimates.

As shown in the sections below, Italy seems to want to spend more on social services than the country can afford, given its present stage of development, while refusing to promptly pay the price in terms of higher taxes. However, even now it is clear that a real strategy for recovery must be aimed at curbing and redistributing expenditure, while the tax structure must be readjusted with a view to reducing the degree of inequality and ensuring that fiscal measures incentivate capital accumulation and growth rather than seriously jeopardizing future prospects, as is currently the case.

BREAKDOWN OF GENERAL GOVERNMENT EXPENDITURE
AND REVENUE, PER CAPITA GDP BEING EQUAL

TABLE 5A

Country	Year	Capital account expenditure	Current account expenditure	Revenue from direct taxes	Revenue from indirect taxes	Revenue from social security contributions
Italy	1974	9.5	90.5	21.3	33.1	41.1
United Kingdom	1972	11.3	88.7	38.0	39.5	14.7
France	1971	10.3	89.7	17.4	38.9	34.3
Germany	1971	11.7	88.3	29.9	34.8	31.9
Japan	1977	22.0	78.0	37.8	28.5	27.9

Source: OECD: *National Accounts*, 1983 and 1989.

TABLE 5B

Country	Year	Capital account expenditure	Current account expenditure	Revenue from direct taxes	Revenue from indirect taxes	Revenue from social security contributions
Italy	1981	10.4	89.6	31.8	24.0	33.2
United Kingdom	1979	8.3	91.7	34.9	38.5	15.2
France	1976-1977	11.4	88.6	18.5	32.0	36.0
Germany	1976-1977	12.1	87.9	29.0	28.1	34.4
Japan	1981	21.4	78.6	39.4	26.4	27.2

Source: OECD: *National Accounts,* 1983 and 1989.

TABLE 5C

Country	Year	Capital account expenditure	Current account expenditure	Revenue from direct taxes	Revenue from indirect taxes	Revenue from social security contributions
Italy	1985	11.6	88.4	33.9	23.4	31.2
United Kingdom	1985	6.4	93.6	36.1	36.7	16.1
France	1980	9.9	90.1	17.9	31.1	37.9
Germany	1978	11.0	89.0	30.1	29.4	32.4
United States	1965	8.3	91.7	47.6	33.7	16.1
Japan	1983	19.0	81.0	39.3	24.9	27.6

Source: OECD: *National Accounts,* 1983 and 1989.

TABLE 5D

Country	Year	Capital account expenditure	Current account expenditure	Revenue from direct taxes	Revenue from indirect taxes	Revenue from social security contributions
Italy	1989	9.4	90.6	34.1	24.8	33.5
United Kingdom	1988	6.3	93.7	33.8	41.1	17.4
France	1987	8.6	91.4	18.9	29.1	37.9
Germany	1985-1986	8.6	91.4	41.4	26.5	22.8
United States	1965	8.3	91.7	47.6	33.7	16.1
Japan	1986-1987	17.8	82.2	39.1	24.6	26.4

Sources: OECD: *National Accounts,* 1983 and 1989; BANK OF ITALY: *Annual Report,* 1990.

1.2 Tax Revenue

The above considerations are confirmed by the data contained in Table 6, an outline of the enormous effort made by the Italian economy to increase tax revenue over the last decade. Expressed in per cent of GDP Italian tax revenue has risen by approximately 10 points, surpassing the British figure and reaching the European average, more than 10 points away from the United States and Japan. Since Italy's overall revenue still lags behind the levels reached in the other countries, the obvious consequence is that a large amount of effort must come from the area of non-tax revenue, mainly the pricing of government services. These were often wrongly used as a device to lessen inflationary indicators. The backlash came in the form of operating losses and bailout programmes for the agencies and firms wholly dependent on government funds; this acted as a disincentive to effective managerial decision making within these organizations, thus undermining management and discouraging the pursuit of informed development strategy.

TABLE 6

TAX REVENUE, IN % OF GDP

Year	Italy	United Kingdom	France	Germany	United Kingdom France Germany	United States	Japan
1980	29.7	35.6	40.8	41.3	39.3	28.4	25.4
1981	30.7	37.7	40.9	41.2	39.9	29.1	26.6
1982	32.7	38.3	41.8	39.6	39.9	28.4	26.9
1983	34.0	38.0	42.5	40.8	40.4	28.0	27.2
1984	33.8	38.0	43.5	41.1	40.9	28.1	27.7
1985	33.9	37.9	43.4	41.4	40.9	28.6	28.3
1986	34.4	37.7	42.9	40.7	40.4	28.6	28.4
1987	35.3	37.6	43.4	40.7	40.6	29.4	29.9
1988	35.8	37.4	43.2	40.6	40.4	29.4	29.9
1989	37.4	37.2	43.0	40.5	40.2	29.4	29.9
1990 (*)	39.4	37.0	42.7	40.4	40.0	29.4	29.6

(*) Estimates.
Source: OECD, 1990; EEC, 1990.

Moreover, in many cases the monopolistic behaviour of these firms and agencies impeded the healthy control exercised by the consumer in a competitive market. Consumer-taxpayers were compelled to pay very little outright and yet to pay much more in taxes for substandard, unreliable public services. Consumers who could afford it often replaced them with private sector substitutes therefore paying three times to actually use a service. Unfairly, those who could not afford private services had to accept what was offered by the public sector.

Other salient aspects arise concerning those responsible for making social security contributions. The data exhibited in Table 7 clearly point to another Italian anomaly, the fact that despite slight corrections in the trend from 1980 to 1987, firms have paid the highest social security contributions of any of the six most industrialized countries: this graph has steadfastly remained at over two-thirds of

TABLE 7

PERCENTAGE BREAKDOWN OF SOCIAL SECURITY CONTRIBUTIONS BY

Country	Employers	Employees	Professionals
1980			
Italy	74.7	18.2	7.1
United Kingdom	59.0	38.6	2.4
France	66.5	26.0	7.5
Germany	53.8	44.5	1.7
United States	59.2	38.2	2.7
Japan	50.9	35.1	14.1
1987			
Italy	70.3	19.4	10.3
United Kingdom	51.7	45.7	2.7
France	63.2	28.6	8.2
Germany	51.2	43.1	5.7
United States	57.8	38.5	3.7
Japan	51.6	35.8	12.7

Source: OECD: *Statistiques Des Recettes Publiques,* 1989.

the total in Italy, as against approximately 55% on average in the other countries.

1.3 Composition of Italian Government Expenditure

A detailed analysis of the line items of revenue inflows and expenditure outflows in Italian government budget from 1970 to the present is provided in Table 8. Of particular note is the percentage composition of current expenditure according to the economic classification, during the 1980s. As shown in Table 9, this grouping makes it possible to determine what is at the root of current expenditure growth. In fact, wages and salaries account for a solid 30% of the total, and social services, mainly pensions, account for 36%. Thus, two-thirds of the current expenditure of general government is a function of the cost of labour (pensions are linked in large part to wages). Further, the purchase of goods and services (for intermediate consumption, such as desks, chairs, pens, inkwells, computers, syringes) amounts to about 11% of expenditure; however, this cannot be reduced significantly.

Finally, there is the vicious deficit-debt-interest-debt cycle that took full flight during the 1980s. Interest paid by government touched 14% of total expenditure in 1980 and was rapidly approaching 20% in 1989-1990. Thus, the tie which in the private sector binds the cost of labour to the conditions for competitiveness on domestic and international markets recurs in the public sector as the link between public expenditure and the cost of labour. Public sector labour contract negotiation are a telling and often negative experience: these renewals swallowed up the increases in revenue raised over the past three years.

1.4 Industrial Sector Subsidies and Taxes

Also deserving of mention is the net amount paid by industry sectors in the form of indirect taxes net of production subsidies and social security contributions. In Table 10, which reveals the inequalities of payment distribution among economic sectors, one figure stands out from the others. Whereas industry, commerce and credit

TABLE 8

INCOME AND OUTLAY ACCOUNT FOR GENERAL GOVERNMENT

	1970	1971	1972	1973	1974	1975	1976	1977	1978	1979
Revenue	*(percentage composition of total revenue)*									
Direct taxes	17.38	17.69	19.57	19.14	18.76	20.55	22.38	24.00	26.96	26.67
Indirect taxes	35.23	33.54	31.43	30.92	30.50	27.07	27.85	28.49	26.95	25.41
Social security contributions (*)	38.28	39.00	39.29	40.59	40.88	44.08	41.99	39.49	38.58	40.52
Other	7.45	8.47	8.60	8.44	8.04	7.33	6.45	6.07	6.81	6.91
Total current revenue	98.34	98.70	98.89	99.09	98.18	99.03	98.66	98.05	99.30	99.51
Total capital revenue	1.66	1.30	1.11	0.91	1.82	0.97	1.34	1.95	0.70	0.49
Total revenue	100.00	100.00	100.00	100.00	100.00	100.00	100.00	100.00	100.00	100.00
Expenditure	*(percentage composition of total expenditure)*									
Interm. public consumption	50.60	52.00	51.41	51.30	47.10	44.61	42.91	43.62	42.35	44.38
Compens. to employees	30.35	31.11	30.60	31.04	29.27	25.88	25.60	26.23	25.47	26.40
Social security	35.61	35.57	36.04	37.07	36.14	35.68	36.20	34.80	35.18	34.29
Production subsidies	4.30	4.88	4.54	4.09	4.16	6.81	6.29	6.71	6.61	6.78
Deprec. net oper. prof.	−1.60	−1.77	−1.43	−1.43	−3.49	−1.22	−1.04	−0.97	−0.77	−0.71
Other transfers	2.30	2.49	2.39	2.88	2.55	2.60	2.62	2.05	2.25	1.69
Curr. exp. net of interest	84.37	84.44	84.30	84.69	81.62	79.48	79.15	78.62	78.04	77.86
Interest payments	5.09	5.25	5.56	6.51	7.76	8.72	10.09	10.70	12.03	12.13
Total current expenditure	86.98	89.69	89.86	91.19	89.38	88.19	89.24	89.32	90.07	90.00
Investment	9.19	8.08	8.03	7.64	8.20	8.14	7.97	7.61	6.75	6.77
Investment subsidies	4.04	2.23	2.11	2.04	2.42	3.66	2.79	3.07	3.19	3.31
Total capital expenditure	13.02	10.31	10.14	9.68	10.62	11.81	10.76	10.68	9.93	10.00
Total expenditure	100.00	100.00	100.00	100.00	100.00	100.00	100.00	100.00	100.00	100.00

(*) Including imputed contributions.
Source: ISTAT: *Conti delle amministrazioni pubbliche*, 1989.

TABLE 8 continued

	1980	1981	1982	1983	1984	1985	1986	1987	1988
Revenue	*(percentage composition of total revenue)*								
Direct taxes	28.76	31.64	32.03	31.59	32.92	33.69	32.60	33.47	33.16
Indirect taxes	25.85	23.80	23.16	23.37	24.23	23.21	23.07	23.85	24.96
Social security contributions (*)	38.17	36.97	37.04	35.80	35.48	35.21	35.40	34.95	34.27
Other	6.45	6.64	5.43	6.09	6.03	7.17	8.28	7.07	6.86
Total current revenue	99.23	99.04	97.65	96.85	98.66	99.28	99.34	99.35	99.25
Total capital revenue	0.77	0.96	2.35	3.15	1.34	0.72	0.66	0.65	0.75
Total revenue	100.00	100.00	100.00	100.00	100.00	100.00	100.00	100.00	100.00
Expenditure	*(percentage compostion of total expenditure)*								
Interm. public consumption	44.35	43.48	42.04	41.96	41.83	41.63	41.09	42.95	43.59
Compens. to employees	26.52	26.45	25.12	24.43	24.14	23.32	22.96	23.96	24.29
Social security	33.62	34.03	33.69	34.66	33.57	33.51	33.73	34.02	33.76
Production subsidies	6.80	6.22	6.42	5.83	6.20	5.51	6.04	5.34	4.85
Deprec. net oper. prof.	−0.78	−0.61	−0.71	−0.74	−0.78	−0.73	−0.71	−0.70	−0.73
Other transfers	1.69	1.31	1.28	1.52	1.56	1.64	1.72	1.76	1.88
Curr. exp net of interest	77.16	76.22	74.62	74.83	73.93	72.76	73.15	74.22	74.07
Interest payments	12.59	13.36	14.77	15.00	16.05	15.68	16.73	15.79	16.12
Total current expenditure	89.75	89.58	89.39	89.83	89.98	88.44	89.88	90.01	90.19
Investment	7.57	7.93	7.75	7.48	7.25	7.31	6.97	6.89	6.76
Investment subsidies	2.69	2.49	2.86	2.69	2.76	4.24	3.16	3.11	3.05
Total capital expenditure	10.25	10.42	10.61	10.17	10.02	11.56	10.12	9.99	9.81
Total expenditure	100.00	100.00	100.00	100.00	100.00	100.00	100.00	100.00	100.00

(*) Including imputed contributions.
Source: ISTAT: *Conti delle amministrazioni pubbliche*. 1989.

TABLE 9

GENERAL GOVERNMENT EXPENDITURE
(on a cash basis, L bn at current prices)

	1980		1985		1989	
	absolute values	(%)	absolute values	(%)	absolute values	(%)
Total current expenditure	150,854	100	383,502	100	580,845	100
of which:						
Compensation to employees	49,073	32.5	111,084	29.0	163,903	28.2
Intermediate consumption......	17,171	11.4	45,247	11.8	65,003	11.2
Social security	54,696	36.3	139,055	36.2	210,803	36.3
Production subsidies	6,255	4.1	14,168	3.7	19,417	3.4
Interest payments	20,634	13.7	66,352	17.3	108,109	18.6
Other	3,025	2.0	7,596	2.0	13,610	2.3

Source: BANK OF ITALY: *Annual Report*, various years.

take on a heavy burden, agriculture and transport make a negative contribution, that is, these sectors receive net subsidies which in 1988 amounted to 7% of value added for agriculture and 9% of the same figure for transport.

TABLE 10

INDIRECT TAXES NET OF PRODUCTION SUBSIDIES
AND SOCIAL SECURITY CONTRIBUTIONS, IN % OF VALUE ADDED

	1980			1988		
	Indirect tax	Soc. sec. contr.	Total	Indirect tax	Soc. sec. contr.	Total
Agricolture............	−5.8	2.1	−3.7	−8.9	2.2	−6.7
Industry	4.9	14.4	19.3	7.8	14.3	22.1
— Energy	46.9	7.9	54.8	49.1	6.8	55.9
— Process.	0.7	16.1	16.7	0.9	16.5	17.4
— Constr.	−1.0	11.4	10.4	−0.1	11.7	11.5
Services	−2.2	10.0	7.7	−1.0	9.3	8.3
— Comm.	−0.8	6.8	6.0	0.1	7.1	7.2
— Trasp.	−36.2	20.2	−15.9	−26.2	16.8	−9.4
— Credit	8.0	26.1	34.1	9.3	21.1	30.4
— Other	4.6	4.8	9.4	3.6	5.9	9.5
Non-mkt serv.	0.0	23.3	23.3	0.0	24.6	24.6
Total	0.1	1.3	1.3	1.8	12.8	14.6

Source: ISTAT: *Conti delle amministrazioni pubbliche*, 1989.

1.5 *Who Receives Government Expenditure*

In Table 11, we regrouped line items of public sector expenditure received by households and by firms (1).

As shown in the Table, during the 1980s the average amount received by households stood at 70% of the total government expenditure, while the amount received by firms was equal to about 30%. Table 11 also shows that in recent years the amount of government expenditure received by firms has fallen sharply (−2.5 points from 1985-1989) in favour of public expenditure received by households (+2.5 points from 1985-1989).

1.6 *Breakdown of Savings and Financial Balances in Italy and in the Major Industrialized Countries*

Each sector of the economic system is both a recipient of income and an expenditure decision making centre. The balance of current revenue and expenditure within one sector is defined as "savings", positive in the case of a surplus, and negative in the case of a deficit. Negative savings within one sector, then, mean that insufficient revenues are generated to cover the current expenditure on consumption. For a household, this would mean taking a bank loan to cover everyday expenses; for a firm, one example might be paying wages and salaries by means of current account overdrafts or mortgages.

Data concerning the composition of savings normally make reference to four institutional sectors: firms, financial institutions, public sector and households. The algebraic sum of these four flows make up national savings. The current account portion of the balance of payments is added to determine the total savings available within the economy. The amount by which the current account balance is positive is equivalent to the portion of national savings that has been absorbed abroad, and is thus inaccessible to the domestic system. Conversely, the amount by which the balance is negative is equivalent

---

(1) Interest payments were divided between households and firms by calculating the percentage of public debt held by households in proportion to total public debt net of debts to Bank of Italy-Foreign Exchange Office (UIC).

GOVERNMENT EXPENDITURE TO HOUSEHOLDS AND FIRMS
(L bn, at current prices)

	1980	(%)	1985	(%)	1988	(%)	1989	(%)
Households	109,286	69.3	278,655	68.9	383,810	71.1	428,835	71.5
compensation	49,073	31.1	111,084	27.5	151,286	28.0	163,903	27.3
transfers	54,696	34.7	139,055	34.4	189,532	35.1	210,803	35.1
interest	5,517	3.5	28,516	7.1	42,992	8.0	54,129	9.0
Firms	48,475	30.7	125,619	31.1	156,055	28.9	171,269	28.5
goods and services	17,171	10.9	45,247	11.2	61,201	11.3	65,003	10.8
production subsidies	6,255	4.0	14,168	3.5	16,808	3.1	19,417	3.2
investment subsidies	3,365	2.1	7,388	1.8	9,167	1.7	9,148	1.5
investment	13,856	8.8	34,877	8.6	45,177	8.4	50,551	8.4
shares and partic.	4,069	2.6	10,835	2.7	12,786	2.4	12,786	2.1
other	3,759	2.4	13,104	3.2	10,916	2.0	14,364	2.4
Total	157,761		404,274		539,860		600,104	

Sources: Relazione generale sulla situazione economica del paese, various years; Bank of Italy Annual Report, various years; Conto riassuntivo del Tesoro, various years; our estimates.

to the portion of savings generated abroad that is used by the national economy. Thus defined, total savings refer to the real resources available for investment.

As shown in Table 12, two Italian institutional sectors consistently show positive savings: households and financial institutions, epitomizing what has often been called one of the cornerstones of the Italian economy, the massive flow of household savings. Though somewhat lower than the levels achieved by households, savings held by financial institutions are nonetheless impressive. The latter began to rise steadily in the 1970s and remained consistently high from the mid-1980s onward.

A glaring inconsistency is revealed among non-financial institutions, which for almost ten years have borne the brunt of the serious financial and economic difficulties of the system. During 1974-1983,

TABLE 12

BREAKDOWN OF SAVINGS IN ITALY

	Non-financial institutions	Financial institutions	General government	Households	External sector (*)	Total available savings
1970	395	514	116	9,079	−761	10,873
1971	15	512	−1,381	10,876	−1,570	11,601
1972	161	515	−3,113	12,822	−1,168	11,564
1973	355	670	−3,544	15,088	+1,472	11,109
1974	−698	1,356	−4,211	17,646	+5,212	8,900
1975	−4,998	2,477	−8,862	23,580	+377	11,843
1976	−4,286	2,923	−8,007	28,232	+2,343	16,556
1977	−5,173	3,432	−8,115	33,229	−2,175	25,604
1978	−6,413	4,739	−12,621	41,589	−5,266	32,637
1979	−2,637	5,968	−14,365	46,315	−4,552	39,969
1980	−2,830	8,733	−11,959	48,591	+8,291	34,671
1983	−112	8,570	−44,511	96,650	−2,030	62,617
1984	1,467	10,843	−53,532	113,955	+4,662	65,137
1985	11,598	13,330	−58,253	107,930	+7,459	67,146
1986	28,142	16,700	−62,744	98,072	−4,290	84,460
1987	32,330	11,479	−64,586	104,363	+2,167	81,419

Source: BANK OF ITALY: *Annual Report*, various years; OECD: *National Accounts*, 1990.
(*) (−) Current account surplus, (+) Deficit.

firms held negative savings which soared toward the end of the 1970s. Only in 1984 did things return to "normal", that is, savings were once again positive, and remained so in subsequent years. These data show the positive effects of the restructuring which took place over the last decade and which restored financial and economic equilibrium to firms. To be sure, the flows of positive savings during the late 1980s seem massive compared with the negative situation of the 1970s. However, the conditions of the 1970s were an aberration and could scarcely be maintained for very long; the phenomenon of recent years, merely brought Italy in line with the other leading industrialized countries.

This return to normal experienced by firms did not occur in the case of the public sector. Herein lie the cause of the progressive worsening of Italian public finances which has continued for twenty years with no clear sign of change.

Until 1970, in fact, the Italian public sector made a positive contribution to national savings. This meant that the Italian public debt was entirely a consequence of capital account expenditure which was partially offset, albeit to a limited extent, by a current account surplus in the government budget. However, as of 1971 general government revenues were no longer sufficient to cover even total current expenditure. Aside from a brief slowdown in 1980, this grew, depleting national savings to the tune of L70 trillion a year for the past three years. In the next section, we will attempt to estimate the structural impact of Italian public finance on the levels of productive capacity and full employment of the economy.

The breakdown of Italian savings is compared with those of the other industrialized countries in Table 13, and values are expressed in per cent of GDP. Of particular importance are the positive and negative differences of Italy's situation. As shown earlier, only Italy and Japan remain steadily and sometimes well above 15% of GDP. The other industrialized countries all show levels of household savings lower, and in some cases far lower, than 10% of each nation's GDP. On the other hand, whereas in Japan the rest of the economy, including the public sector, contribute positively to national savings, in Italy the massive flow of household savings is flanked by the negative savings of general government and, up to 1984, of firms as well. This is the peculiar trait which reveals the weakness of Italy's financial and economic systems. For several years, this characteristic was apparent in Great Britain, and increasingly during the 1980s, in the United States. Furthermore, the sizeable savings generated by financial institutions which throughout the 1970s served to counterbalance the negative position of firms, reveals an uneven redistribution of resources from industry to banks. The accumulation of savings within the banking and financial system appears particularly striking when compared with the other industrialized countries. The opposite trend of recent years, according to which the savings of Italian financial institutions fell to 2% of GDP in 1986 and a low of about 1% in 1987, should not be underestimated.

This sharp decline is linked to the economic and financial redress of firms which led to a lower degree of intermediation by the Italian banking sector and more industrial sector indebtedness on international markets. However, this could also explain a state of abnormal concentration and insufficient competitiveness which has marked the Italian banking system, a state which this sector can ill afford to maintain as the market begins to obey normal rules of competition. It becomes a question of accepting the challenge of increased efficiency and competitive survival facing the entire Italian financial system.

Overall, the situation of the Italian economy rests squarely on household savings, which despite the havoc wrought elsewhere, especially in government, permitted national savings to remain well above 15% of GDP almost until 1984. Particularly worrisome is the trend of recent years toward a lowering of household savings to about 11% compared with a high of 19% reached in 1982. However, due caution must be exercised in evaluating these reductions, since they may derive at least in part from accounting phenomena, such as the exclusion of expenditure on durable goods, or investments in human capital (education, longer training periods, and so on), in the definition of savings (2).

---

(2) An attempt to reformulate the Life Cycle model to introduce investments in human capital and choices between working hours and leisure time to explain the reduction in savings in terms of a rebound effect producing repercussions in subsequent periods was recently proposed in: BALDASSARRI M. *et* AL. [5]. This paper introduces additional demographic elements (such as the aging of the population) widely used to interpret the reduced propensity to save in industrialized countries.

TABLE 13

BREAKDOWN OF SAVINGS IN SELECTED COUNTRIES
(in % of GDP)

Country	1970	1974	1978	1970	1974	1978
	Total savings			*Households*		
Italy	14.9	17.3	9.8	14.8	16.3	18.7
United Kingdom	11.6	11.0	7.8	4.1	5.3	6.3
France	15.6	16.4	10.7	8.8	10.0	10.2
Germany	17.8	11.7	10.7	11.8	9.3	8.5
United States	7.2	7.5	7.9	6.1	6.4	3.9
Japan	25.7	23.9	17.5	11.6	17.1	15.2
	Financial institutions			*General government*		
Italy	0.8	1.2	2.1	0.2	−3.8	−5.7
United Kingdom	0.5	1.0	1.0	7.5	1.0	−1.7
France	0.6	0.9	1.0	4.3	3.6	0.3
Germany	0.9	1.4	1.4	5.9	4.0	2.0
United States	0.5	0.5	0.7	0.0	0.7	0.8
Japan	1.6	1.7	0.6	6.7	6.3	1.4
	Firms			*External sector*		
Italy	0.3	−1.0	−2.9	−1.2	4.6	−2.4
United Kingdom	0.8	−0.9	2.6	−1.3	4.6	−0.4
France	2.0	0.4	−0.2	−0.1	2.3	−0.6
Germany	−0.2	−0.4	0.2	−0.6	−2.6	−1.4
United States	0.8	0.2	1.9	−0.2	−0.3	−0.6
Japan	6.8	−2.2	2.0	−1.0	1.0	−1.7

Source: OECD: *National Accounts*, 1981, 1990 and our estimates.

TABLE 13 continued

Country	1980	1981	1982	1983	1984	1985	1986	1987	1988
					Total savings				
Italy............	9.1	6.7	8.3	9.9	9.0	8.3	9.4	8.3	n.a.
United Kingdom	7.6	7.5	7.0	6.8	5.3	7.0	4.1	3.7	n.a.
France	10.6	7.6	4.9	5.4	6.1	6.3	7.8	6.4	7.4
Germany	10.3	10.8	11.1	10.9	12.7	14.2	15.9	15.5	12.7
United Kingdom	6.4	6.7	2.7	1.2	1.9	0.4	−1.0	+1.3	0.5
Japan	17.3	18.4	17.7	17.9	19.9	21.6	22.4	22.0	22.1
					Households				
Italy............	16.0	16.1	19.0	15.3	15.7	13.3	10.9	10.7	n.a.
United Kingdom	6.4	5.7	5.1	4.0	4.1	3.2	1.7	0.2	n.a.
France	8.9	9.6	9.1	7.9	6.8	6.5	5.8	4.4	4.9
Germany	8.3	8.9	8.4	7.0	7.3	7.3	7.7	7.8	7.9
United States	6.4	6.6	6.4	5.4	5.8	4.6	4.5	3.9	4.7
Japan	12.9	13.1	11.8	11.7	11.2	11.0	11.3	10.3	9.9

Source: OECD: *National Accounts*, 1981, 1990 and our estimates.

TABLE 13 continued

Country	1980	1981	1982	1983	1984	1985	1986	1987	1988
Firms									
Italy	0.0	0.0	0.0	0.0	−0.2	1.4	3.1	3.3	n.a.
United Kingdom	−0.4	0.7	0.3	2.1	2.3	2.6	2.2	3.4	n.a.
France	−0.7	−1.6	−1.7	−1.5	−0.8	−0.5	0.8	1.0	1.3
Germany	0.3	0.4	0.0	0.7	1.0	0.9	0.8	1.1	n.d.
United Kingdom	0.8	1.1	0.4	1.5	2.6	2.5	1.8	1.6	1.5
Japan	2.2	1.7	2.0	1.8	2.4	2.6	3.2	3.1	2.8
Financial institutions									
Italy	0.0	0.0	0.0	1.4	1.5	1.6	1.9	1.2	n.a.
United Kingdom	1.5	1.5	1.4	1.4	1.2	1.8	2.2	2.2	0.0
France	0.8	0.4	0.3	1.1	1.3	1.5	1.8	1.8	1.4
Germany	1.6	1.8	1.7	1.8	1.7	1.5	1.2	1.4	0.0
United States	0.3	0.1	0.0	0.3	−0.1	0.1	0.2	0.1	0.0
Japan	0.6	0.1	0.4	0.2	0.1	0.0	−0.7	−1.0	−1.0

Source: OECD: *National Accounts,* 1981, 1990 and our estimates.

TABLE 13 continued

Country	1980	1981	1982	1983	1984	1985	1986	1987	1988
General government									
Italy	−4.7	−7.2	−8.7	−7.0	−7.4	−7.2	−7.0	−6.6	−6.2
United Kingdom	−1.6	−1.6	−1.4	−1.6	−2.1	−1.2	−1.2	−0.5	n.a.
France	2.2	0.1	−0.7	−1.4	−1.2	−1.3	−1.1	−0.6	0.2
Germany	1.8	0.5	0.4	0.7	1.4	2.0	1.7	1.1	0.7
United States	−1.6	−1.3	−4.2	−5.0	−3.9	−4.0	−4.3	−3.4	−3.3
Japan	2.6	3.1	2.8	2.3	3.3	4.3	4.2	6.0	7.5
External sector									
Italy	−2.2	−2.2	−1.9	+0.3	−0.6	−0.9	+0.5	−0.2	−0.6
United Kingdom	1.6	2.6	1.5	0.9	−0.2	0.7	−0.9	−1.6	n.a.
France	−0.6	−0.8	−2.1	−0.8	0.0	0.1	0.5	−0.3	−0.4
Germany	−1.8	−0.8	0.5	0.7	1.3	2.6	4.4	4.1	4.1
United States	0.4	0.3	0.0	−1.0	−2.4	−2.9	−3.2	−3.4	−2.4
Japan	−1.0	0.5	0.7	1.8	2.8	3.7	4.3	3.7	2.8

Source: OECD: *National Accounts*, 1981, 1990 and our estimates.

Nonetheless, this phenomenon is worthy of consideration because if Italian households exhibited a strong tendency toward reducing their savings, and if the critical situation of public finances continued to persist, then the development potential of Italian economy could be drastically reduced.

The total balances, or financial balances, investment expenditure included, of each institution are examined in Table 14. In this case, total income of each sector is compared with total expenditure, consumption expenditure and investment expenditure.

If revenue exceeds expenditure, then a financial surplus has been achieved; if expenditure exceeds revenue, then the result is described as the total borrowing requirement. From this perspective as well, households prove to be the driving force underlying this process. The total surplus of this sector rose from about L9 trillion in 1971 to about L114 trillion in 1988. This positive balance was invariably ac-

TABLE 14

FINANCIAL BALANCES
(L bn)

	Households	Firms	Public sector	External sector
1971	8,779	− 3,577	− 4,892	− 981
1972	10,292	− 2,561	− 6,791	− 1,169
1973	14,364	− 6,972	− 9,114	1,473
1974	14,052	− 8,757	− 10,827	5,212
1975	20,726	− 9,686	− 15,566	377
1976	22,745	− 11,804	− 14,459	2,343
1977	28,089	− 9,673	− 17,051	− 2,175
1978	34,753	− 5,946	− 23,719	− 5,260
1979	39,258	− 3,558	− 26,989	− 4,552
1980	45,844	− 29,239	− 33,559	8,532
1981	59,075	− 27,991	− 52,019	10,301
1982	73,365	− 22,335	− 66,496	8,432
1983	92,746	− 27,638	− 76,515	− 2,323
1984	102,483	− 25,273	− 94,766	4,314
1985	111,059	− 35,898	− 114,048	7,102
1986	120,683	− 18,070	− 108,794	− 3,802
1987	131,637	− 31,650	− 115,503	1,940
1988	152,892	− 50,318	− 126,190	7,819
1989	143,900	− 55,900	− 130,100	14,500

Source: BANK OF ITALY: *Annual Report*, various years.

companied by the borrowing requirements of firms and the public sector; in 1989, these figures equalled L56 trillion and L130 trillion respectively. Except for 1983 and 1986, the external sector played a significant role in the financing of the Italian economy, especially during the 1980s. Its contribution amounted to L15 trillion and exceeded 1% of GDP in 1989 as a result of massive flows of foreign capital toward the Italian economy. The peculiar nature of this system is shown in Table 15, which compares it once again with the other industrialized economies. The household positive balance is by far the highest in the world, and remains consistently above 10% of GDP. The

TABLE 15

FINANCIAL BALANCES IN SELECTED COUNTRIES
(in % of GDP)

Country	1970	1974	1978	1970	1974	1978
	\multicolumn{3}{c}{Households}	\multicolumn{3}{c}{Firms}
Italy	9.8	12.1	15.2	−5.4	−10.8	−5.0
United Kingdom	2.9	5.1	5.6	−2.5	−5.8	0.0
France	3.7	3.7	5.2	−4.7	−6.8	−3.4
Germany	7.9	8.6	6.9	−8.1	−5.4	−3.7
United States	3.2	3.7	0.2	−2.5	−4.5	−2.3
Japan	6.8	9.9	9.7	−8.6	−12.4	−2.8
	\multicolumn{3}{c}{Financial institutions}	\multicolumn{3}{c}{General government}
Italy	0.2	1.0	1.9	−3.5	−7.0	−9.7
United Kingdom	−0.6	−0.5	−0.5	2.5	−3.8	−4.3
France	−0.1	0.0	0.5	0.9	6.3	−1.8
Germany	0.3	0.7	0.9	0.3	−1.4	−2.7
United States	0.3	0.3	0.5	−0.6	0.4	0.6
Japan	0.9	1.1	0.8	1.8	0.4	−5.9
	\multicolumn{3}{c}{External sector}			
Italy	−1.2	4.6	−2.4			
United Kingdom	−1.3	4.6	−0.4			
France	−0.1	2.3	−0.6			
Germany	−0.6	−2.6	−1.4			
United States	−0.2	−0.3	0.6			
Japan	−1.0	1.0	−1.7			

Source: OECD: National Accounts, 1981, 1990 and our estimates.

TABLE 15 continued

Country	1980	1981	1982	1983	1984	1985	1986	1987	1988
Households									
Italy	11.4	12.8	15.8	14.2	14.4	13.0	12.7	13.3	n.a.
United Kingdom	5.5	4.9	3.8	2.6	2.9	2.0	0.1	−1.6	n.a.
France	3.7	4.9	4.7	4.0	3.6	3.6	2.8	1.6	1.8
Germany	7.3	8.0	7.2	6.0	6.3	6.2	6.6	6.5	n.a.
United States	4.2	4.4	6.2	5.8	4.9	3.3	4.0	3.1	3.7
Japan	7.0	9.8	9.5	9.2	9.3	9.3	9.2	9.1	8.5
Firms									
Italy	−7.0	−4.8	−4.8	−4.4	−3.9	−4.1	−1.6	−2.4	n.a.
United Kingdom	−0.5	1.5	1.2	1.7	1.3	1.4	1.5	1.6	n.a.
France	−5.0	−4.2	−4.4	−2.7	−2.0	−1.8	−1.0	−1.4	−1.8
Germany	−7.4	−6.3	−4.4	−3.8	−4.2	−3.4	−1.7	−1.6	n.a.
United States	−0.9	−1.3	−0.4	−0.3	−1.4	−0.6	−1.1	−0.6	−0.9
Japan	−5.1	−4.7	−6.0	−5.3	−5.6	−4.3	−3.5	−6.1	−7.2
Financial institutions									
Italy	1.8	2.0	2.1	1.3	1.3	2.0	1.8	0.9	n.a.
United Kingdom	−0.3	−0.3	−0.3	0.3	−0.3	0.3	1.1	1.2	n.a.
France	0.7	0.3	0.2	0.9	1.1	1.1	1.3	1.4	1.0
Germany	1.2	1.3	1.0	1.0	1.0	0.8	0.7	0.7	n.a.
United States	0.6	−0.5	−0.3	−0.2	−0.6	−0.7	−0.4	−0.7	−1.1
Japan	0.9	−0.2	0.6	1.7	1.4	−0.2	−0.5	−0.7	0.7
General government									
Italy	−7.2	−13.3	−12.8	−10.9	−11.9	−12.8	−11.2	−11.0	n.a.
United Kingdom	−3.6	−4.0	−2.9	−3.3	−3.9	−2.7	−2.8	−1.4	n.a.
France	0.0	−1.9	−2.8	−3.2	−2.8	−2.9	−2.7	−2.0	−1.4
Germany	−2.9	−3.7	−3.3	−2.5	−1.9	−1.1	−1.3	−1.8	n.a.
United States	−2.4	−2.9	−5.1	−5.6	−5.0	−5.1	−5.3	−4.3	−4.0
Japan	−3.9	−4.4	−3.5	−3.8	−2.2	−1.2	−0.9	1.3	2.3
External sector									
Italy	−2.2	−2.2	−1.9	0.4	−0.6	−0.9	−0.4	−0.2	−0.6
United Kingdom	1.1	2.2	1.9	1.2	0.0	1.0	−0.1	−0.3	n.a.
France	−0.7	−0.9	−2.2	−0.9	−1.1	−0.0	0.4	−0.4	0.4
Germany	−1.8	−0.7	0.5	0.7	1.1	2.4	4.2	3.8	n.a.
United Stated	1.2	0.9	1.0	−0.8	−1.9	−2.6	−3.2	−3.4	−2.8
Japan	−1.1	0.4	0.7	1.8	2.8	3.6	4.3	3.7	2.8

Source: OECD: *National Accounts*, 1981, 1990 and our estimates.

other five leading countries never exceed that level, settling on average at about 5%. The other side of the coin is the public sector borrowing requirement, which also remains well above 10% of GDP for all years, while in the other countries it is consistently below 4%, and actually shows a surplus in Japan. Yet another difference, although a less obvious one, between Italy and the other countries is the fact that both the borrowing requirements of firms and the surplus shown by financial institutions are much higher. This confirms the reasoning presented earlier, when discussing the composition of savings, with regard to the atypical relationship between banks and firms within the Italian economy.

1.7 Total Domestic Lending to the Public and Private Sectors

Throughout the 1970s and 1980s, the pace of total domestic lending in Italy accelerated rapidly (Table 16). With this rapid expansion went a veritable reversal of trends concerning the amount of credit absorbed by the private and public sectors. In 1970, approximately two-thirds of total credit was extended to the private sector, while the public sector absorbed the remaining one-third; in 1982, exactly the opposite was true, and in 1983 an unprecedented 70% of total domestic lending was absorbed by Italy's public sector. Even though this trend has corrected itself somewhat in recent years, just under 50% of total domestic credit extended to the private sector, while the public sector continues to account for more than half. A fundamental aspect of the public sector's role in Italy clearly emerges: in real terms the public sector depletes savings by transforming them into consumption, thereby impeding investment, and in financial terms it limits the amount of lending to the private sector by redirecting resources available toward its own consumption expenditure. As shown above, this consists mainly of wages and salaries, pensions and transfer payments to households.

Within this context, it is true that real and financial unbalances were counterbalanced by the behaviour of households, which made it possible to maintain a relatively high level of total savings. However, it

TABLE 16

BREAKDOWN OF TOTAL DOMESTIC LENDING

Year	Absolute values (L bn. at current prices)			Percentage compostion			Percentage of GDP		
	private	state-sector	totale	private	state-sector	total	private	state-sector	total
1970	5,271	2,800	8,071	65.3	34.7	100.0	9.1	4.8	13.9
1973	10,197	10,700	20,897	48.8	51.2	100.0	12.3	13.0	25.3
1977	15,137	20,037	35,174	43.0	57.0	100.0	8.0	10.5	18.5
1980	31,362	34,015	65,377	48.0	52.0	100.0	8.1	8.8	16.9
1981	29,567	45,239	74,806	39.5	60.5	100.0	6.4	9.7	16.1
1982	31,666	69,133	100,799	31.4	68.6	100.0	7.0	15.2	22.2
1983	36,076	85,197	121,273	29.7	70.3	100.0	5.7	13.4	19.1
1984	53,443	91,708	145,151	36.8	63.2	100.0	7.3	12.6	20.0
1985	46,168	107,267	153,435	30.1	69.9	100.0	5.7	13.2	18.9
1986	45,967	106,710	152,677	30.1	69.9	100.0	5.1	11.9	17.0
1987	46,119	105,872	151,991	30.3	69.7	100.0	4.7	10.8	15.5
1988	78,181	118,990	197,171	39.7	60.3	100.0	7.2	11.0	18.2
1989	106,506	122,445	228,951	46.5	53.5	100.0	9.0	10.3	19.3

Source: BANK OF ITALY: *Annual Report*, various years.

is certainly true that household income was constantly supplemented by means of public sector expenditure. In order to explain the complex phenomena involved, one should also take into account the interrelationship which exists among the various segments of the Italian economic system.

2. - The Role of the Government Budget in Supporting Industrial Activity and Long Term Effects on Resource Allocation and on Levels of Productive Capacity and Full Employment

In this section, we have sought to evaluate the effects of the government budget on the Italian economy. To be sure, this type of analysis, based on simulations of econometric models, is subject to the limits of the very structure of the models used; further, it relies on the methodology used when carrying out the simulation.

Bearing in mind these limits, we analyzed firstly the short-term impact of the government budget on the Italian economy, evaluated as an incentive to demand, and thus as an incentive to industry and/or as a determinant of price changes. This could be defined as the most traditional Keynesian approach in which the effects of the level and of the composition of the government budget are measured by means of the multiplier effects on national income. It should be stressed at the outset that the effects on economic growth operate under conditions of flexibility of response of the industrial system which Keynes has and would have defined as availability of unutilized productive capacity, which should maintain steady prices and have a completely real impact on the level of industrial activity. Also deserving of mention is the different impact of differing levels and compositions of public sector revenue and expenditure. The study by Haavelmo makes precise reference to the incorrectness of focusing on "budget balances", whether deficits or surpluses, rather than considering the levels and structure of the various government budget items. It can scarcely be considered surprising, or anti-Keynesian, to witness smaller impact of the government budget on levels of industrial activity and greater impact on prices. On the other hand, the government budget works in concert with the other sectors toward resource allocation and accumulation, causing medium and long-term changes (either positive or negative) in the process of achieving gains in productive capacity and full employment. The role of the public sector in Italy and in the other countries concerning the creation of savings and the allocation of financial resources have already been outlined in the first section. Secondly, therefore, we attempted to assess the structural role of the government budget in Italy throughout the 1970s and 1980s. In many cases, a difficult choice existed between the target of equal distribution and the target of economic growth. Evidence, however, increasingly points to the fact that pursuing equal distribution without taking into account the available real resources, can even produce adverse effects in the parts of society that it seeks to protect. It is thus necessary to take into account a budget constraint matching objectives to resources in the medium and long terms.

Nonetheless, it cannot be doubted that development without equality, or a distorted equality in which all of society is equally impoverished (as in the case of planned economies, or the so-called Chilean "success" story of the 1970s), poses the same questions from a different perspective.

In this sense, the analysis of the short-term impact and of the structural effects of the government budget must result in a more complete and accurate evaluation of the role of the public sector within the economy.

2.1 *Estimates and Evaluations of the Short Term Effects of the Government Budget on the Economic Activity*

This paragraph presents several estimates of the effects of the government budget on production levels (and the indirect effects on employment levels) and on inflation in the short term. The three simulations outlined below cover a wide timespan and make reference to the results presented in a previous paper (3).

We considered it appropriate to make reference to these simulations because they target a period central to the twenty years or so

---

(3) See BALDASSARRI M. [3] and [4].

studied in this paper, more specifically, the years of transition from the realities of the 1970s to those of the 1980s. As will be better demonstrated below, this was the most significant period in which the country wished to pursue a wide range of objectives well beyond the resources available, and thus laid the true groundwork for the current public finance disequilibria. Social reform, state entrepreneurship and support for private sector industry led to burgeoning growth of public expenditure, only partially compensated by the increase in revenue brought about by tax reform and inflation (fiscal drag).

The three simulations were all produced using the same econometric model of the Associazione Prometeia in Bologna. It would thus be useful to complete the study by means of simulations which cover the entire ten year period of the 1980s, even if the results presented here reveal interesting developments.

The three tests are based on the simulation of the impact produced by a 10% change in each expenditure line item, excluding those of the financial kind. Thus an increase of 10% in the government budget deficit is achieved. The results obtained depend on the structural composition of revenue and expenditure line items during each of the budget years studied (some indication of this is given in Table 17) and on the multiplier effects of these same budget items on national income (Table 18). Test results are provided in Table 19.

The impact of the government budget on levels of gross domestic product is very limited during the first year of each test and is then totally absorbed by the negative effect brought about during the subsequent year. Over a span of several quarters, therefore, the overall effect on real income levels and thus on employment levels is virtually zero. It may be interesting to note the small net positive result of the third test for 1982-1983, as a consequence of a structural change in expenditure during this period. A look at the simulations of subsequent years, when these are made available, would also be useful.

Of particular significance is the effect on the rate of inflation measured by the GDP deflator and the consumption price level.

The size and the structure of the government budget during the years studied do not seem to produce the improvement of economic conditions, level of income and employment, traditionally linked to a

TABLE 17

MULTIPLIER EFFECT ON REAL NATIONAL INCOME

Line item	Quarter I	Quarter IV	Quarter VII
Direct taxes	−0.22	−0.56	−0.94
Indirect taxes	−0.27	−0.76	−1.61
Indirect taxes on imports	−0.25	−0.59	−1.27
Employee social-security contributions	−0.22	−0.56	−0.94
Employer social-security contributions	−0.22	−0.71	−1.71
Compensation	1.16	1.00	1.00
Purchase of goods and services	1.16	1.10	1.40
Transfers to households	0.22	0.56	0.94
Transfers to firms	0.27	0.76	1.61
Interest payments	0.22	0.56	0.94
Investment	1.24	1.20	1.40
Current account transfers	0.22	0.10	0.27

Source: CAVAZZUTI F.: *Il nodo della finanza pubblica*, Milano, Feltrinelli, 1979.

TABLE 18

EFFECTS OF 10% CHANGE OF ALL LINE ITEMS, EXCLUDING FINANCIAL ITEMS
(percentage change calculated on real values)

	1st test 1976	1st test 1977	1st test 1978	2nd test 1980	2nd test 1981	3rd test 1982	3rd test 1983
Gross domestic product	0.6	−0.3	−1.0	0.4	−0.4	0.4	−0.2
Household consumption	0.5	—	−0.9	0.2	−0.4	−0.5	−0.2
Total investments and purchases	3.5	3.2	3.0	1.7	1.3	3.3	−1.0
Exports of goods and services	−0.7	−1.4	−1.5	−0.7	−1.1	−1.3	0.7
Imports of goods and services	1.4	2.8	3.2	0.7	1.8	1.1	−2.1
Household disposable income	1.5	—	−0.5	0.3	−0.6	2.4	0.3
GDP deflator	1.2	2.5	3.1	1.3	1.8	4.7	1.2
Household consumption deflator	1.7	3.4	3.8	1.4	2.1	2.7	0.3
Investment deflator	0.7	1.6	1.9	1.3	1.5	1.0	1.0
Export deflator	0.2	0.8	0.9	0.3	0.6	1.2	−0.4

TABLE 19

STRUCTURAL EFFECTS OF THE PUBLIC SECTOR BUDGET
ON POTENTIAL EMPLOYMENT LEVELS
(public sector excluded)

Year	New CSC simulation	Naive test	Prometeia simulation	Previous CSC simulation	New CSC simulation
1988	− 205,000				
1987	− 208,000				
1986	238,000				
1985	− 256,000				
1984	− 262,000				
1983	− 176,000	− 180,000	− 170,000	− 150,000	− 176,000
1982	− 119,000	− 179,000	− 150,000	− 70,000	− 119,000
1981	− 137,000	− 178,000	− 48,000	− 90,000	− 137,000
1980	− 140,000	− 162,000	− 76,000	− 110,000	− 140,000
1979	− 114,000	− 173,000	− 50,000	− 72,000	− 114,000
1978	− 65,000	− 135,000	− 48,000	− 91,000	− 65,000
1977	− 40,000	− 126,000	− 20,000	− 82,000	− 40,000
1976	− 2,000	− 78,000	− 40,000	− 75,000	− 2,000
1975	− 86,000				
1974	− 64,000				
1973	− 36,000				
1972	+ 13,000				
1971	+ 41,000				
Cumulative total	− 2,094,000	− 1,210,402	− 602,000	− 740,000	− 793,000

rise in government expenditure and deficit. In fact, the structure of expenditure and tax levies produce effects which cancel each other out. However, this does not imply that the government budget may be described as being "neutral". As regards the economic conditions, there is actually a sizeable inflationary pressure; as regards the structure of the economy, higher stakes in national output came progressively under government control, concentrating resources within the public sector — a change capable of reshaping even a "mixed" market economy.

2.2 Structural Effects of the Government Budget on Resource Allocation and on Levels of Productive Capacity and Full Employment

After having examined, under the previous heading, the short-term effects on the economy, we will now attempt to measure, as a general tendency, the potential structural effects that the government budget may have produced on the Italian economy in the medium and long term.

The premises underlying this attempt are as follows: *a)* taxes and social security contributions lower income levels of households and firms, thus reducing their ability to accumulate savings and investment; *b)* even limited government expenditure increases investment flows, and acts as a direct incentive to private sector investment; *c)* these effects can be measured jointly in order to evaluate the role of the government budget in the accumulation of capital for industry and, therefore, in raising medium and long-term levels of productive capacity and full employment in Italy.

The first section showed that contrary to the experiences of previous decades, in the early 1970s the savings of the public sector were increasingly negative. What occurred within our economic system was an accelerated distortion in the resource accumulation process. Expressed another way, this means that for every hundred lire of income in taxes, for example, these conditions cause a greater percentage reduction in savings and investment compared with the rise in savings and investment caused by changing these one hundred lire of taxes into public expenditure. Overall, therefore, the government budget produces a small effect on investments which is still relatively pronounced in cases such as Italy's, where government expenditure far exceeds taxes. At the beginning of the next year this decreased flow of investment will translate into reduced productive capacity and lower levels of employment. This phenomenon, seen in the short term (from year to year, for instance) may not appear to be quantitatively significant. However, if this effect is produced consistently over time, these steady (though small) reductions would gradually chip away at nationwide productive capacity and employment, producing a cumulative effect that would be quantitatively significant in the long run.

Moreover, correcting the imbalances of the government budget will become increasingly necessary as time goes by, and increasingly more "politically" awkward. The situation becomes even more complicated as the stock of accumulated public debt activates the vicious cycle of interest on debt, expenditure, deficit, and new debt. Short cuts such as consolidated debt in presence of a persistent budget deficit or reduction of interest rates, heedless of international conditions and of the internal requirement to settle the debt, are not worth discussing. Undoubtedly, a serious problem remains regarding the efficient management of the debt with a view to minimizing costs of servicing the debt load given the internal and international financial constraints.

An initial attempt to measure the structural effects of public budget on the Italian economy was presented during the mid-1980s (4). This study is based on the same type of simulation, obtained using the Centro Studi Confindustria model, and applied to a longer period of time, 1971-1988. The three tests presented earlier referred to 1976-1983. Table 19 summarizes the results of this new simulation and of the previous estimates.

What was defined as the "Naive Test" was carried out with broad evaluations (worked out manually) on the basis of the following assumptions for 1976-83: *a)* a household propensity to save of 15%; *b)* a propensity toward investment of firms of 80% of the revenue of the firm; *c)* a capital/product ratio of 3; *d)* a product/employment ratio of 100 million lire.

The other two tests presented previously were carried out for 1976-1983 using the simulation of two different econometric models developed by the Associazione Prometeia in Bologna and by the Centro Studi Confindustria respectively. In the simulations it was assumed that the current account deficit of the government budget was equal to zero, and thus in the position of "non negative" government savings of the 1950s and 1960s. It was possible for the new simulation to examine a wider timespan, from 1971 to 1988; this simulation was carried out using the Centro Studi Confindustria model.

---

(4) See BALDASSARRI M. [4].

The three simulations refer to a "hypothetical" reduction to zero of the current account balance on the government budget, while the "Naive test" assumes the even less likely case of a total government budget balance equal to zero. For this reason, therefore, while the results of the econometric simulations seem very similar to one another and coherent regarding the two periods examined, the calculations of the "Naive test" show a reduction of productive capacity and full employment levels that is far greater.

As shown in the summarized results of Table 19 (lower employment produced over time by the distortionary effect of the government budget on resurce allocation and on productive capacity), the structural impact continues to increase and assumes significance in quantitative terms.

The previous tests made it possible to estimate the lowering of full employment levels by approximately 600,000 (in the Prometeia simulation) and by approximately 740,000 (in the CSC simulation) as a results of the government current account deficit for 1976-1983.

The new simulation for the same period (Table 19, column 5) confirms the previous figures. The same simulation for the larger timespan of 1971-1988, clearly provides a wider range of these effects, focusing on the gradual nature of this phenomenon. It is also true that during 1976-1983 this negative effect of lowering full employment by 700,000 must be compared with the increase in public sector employment, which during the same period equalled 600,000. However, even in 1983 the net result in terms of total employment was equal to or less than zero: the number of potential jobs eliminated was equal to or higher than the number of jobs created within the public sector. The new estimate proceeds from these results, and permits us to set these effects over nearly twenty years at about two million jobs. During the same period, public sector employment increased by slightly more than 1,400,000. By 1988 the net effect was relevant, showing a loss of 600,000 potential jobs. Small wonder, then, that this trend accelerates toward the end of the 1980s, in concert with the vicious cycle of debt, interest, expenditure, deficit, and more debt.

Also worthy of note is a positive cycle which became much more evident in the new simulation. If the curtailment of public expenditure had occurred at the beginning of the period, the recovery may have been smaller than was considered necessary in subsequent years, and could have had a favorable impact on potential growth, which, in turn, would have generated higher revenue inflows. During sebsequent years, therefore, these developments would have reduced the necessity to cut expenditure to correct current account imbalances and/or would have permitted a reduction of the tax burden. This was the assumption which we introduced into the new simulation. This confirms yet again the importance of when intervention to restore equilibrium occurs and its size and effectiveness in the light of current trends. It is therefore obvious, but deserving of mention, that postponing policy measures increases the amount of correction required. Given the structural developments which come into play, both domestically and internationally, it is not entirely correct to compare time periods which are very distant from one another. However, the period of rapid adjustment of the Italian economy during 1963-1964 should be borne in mind. Whereas it is true that favourable international conditions greatly helped our economy to a balanced growth, it is also true that during the second half of the 1980s the Italian economy was only partially able to profit from favourable international conditions and lost an opportunity to achieve permanent balance, especially as public finances were concerned. In fact, the stock of public debt and the constraints which it poses to the Italian economy have their origins in the delay of the adjustment process and are manifest today in the vicious circle of interest, expenditure and deficit mentioned earlier. In large part, therefore, the debt has become a filter, or the extenuanting feature, of that unfinished portion of restructuring, or more explicity, of the restructuring as yet to be done.

In our view, the various analyses in the previous essays are not mutually exclusive; rather, they include a variety of interpretation which provides increased and more effectual insight into this phenomenon.

Coupled with the work of the previous sections, the data analyzed in the next and final section of this paper is aimed at focusing on the difficulty of accurately dating or identifying the origins of Italy's public debt.

What we wish to point out is the plain fact that the public sector wished to pursue an array of objectives which was much wider than the amount of resources available. Moreover, in pursuing these targets, in spite of a sizeable increase in tax revenue, the public sector set in motion a deficit spiral affecting not only the total balance but also the current account. As we have attempted to demonstrate above, this method of pursuing objectives beyond the reach of existing resources produces the adverse effect of reducing potentially available resources.

In reality, the initial stage of social reforms in the 1970s was followed by a stage of state entrepreneurship, both directly within public sector firms and indirectly, through support of the private sector.

In our view, therefore, the issue is not so much whether to place the blame on one sector or the other for the current state of public finances, any more than it is to praise either sector for the process of industrial restructuring which occurred. What really matters is that the disparity between targets and resources requires, in the final analysis, that somebody settle the accounts. The accumulation of debt meant that the reckoning would be faced by subsequent generations. Furthermore, this process grew exponentially and progressively caused potential resources to dwindle. What was thrust ten or fifteen years ago into "the long term" is what we are experiencing today.

Therefore, in the third section we will seek to determine who acted as the lender of last resort over the last few years and to what degree the current stock of public debt suggests the absence of such a lender, or the presence of a lender who only seemingly and perversely pays the price.

3. - From Inflation Tax to Guaranteed Yields: the Lender of Last Resort During the 1970s and 1980s

In the 1970s, influences from abroad reshaped the Italian economy, radically redistributing wealth and income as a result of the two oil shocks. These developments worked in concert with massive domestic forces on distribution. Apparently, a violent clash occurred between labour unions and firms, and between the two groups and the government. Collective agreements and cost of living indexation for the labour unions, inflation and exchange rate devaluation for firms, inflation and fiscal drag for the government budget seem to have made it possible to overcome those difficult years without actually appealing to a lender of last resort. On closer examination, however, this role is shown to exist (and necessarilly so) and to be played in large part by household-savers who over the years "paid the price" by incurring heavy real losses in financial wealth.

Table 20 outlines the trend in household financial wealth, broken down according to financial assets. Table 21 presents the nominal and real interest rates for each financial asset. Table 22 lists the estimated earnings and real shortfalls for each year.

In our view, these simple but significant data reveal the following: *a)* households incurred steady losses in wealth from 1972 to 1983 and only after this period, following an upturn in real rates of interest, were they able to achieve real gains. These losses amounted to high percentages of GDP; *b)* if the cumulative value of losses is considered to be the algebraic sum of the various years, then the highest value of total loss was achieved in 1983. From L120 trillion, this amount progressively declined as a consequence of gains achieved in subsequent years, until almost all of the loss was offset in 1989 and 1990. However, it is obvious that this calculation should be made for homogeneous and comparable values, at 1990 prices. We therefore re-evaluted, year over year, the shortfall realized in proportion to the rate of inflation. It is therefore clearly evident that the loss of financial wealth was actually much higher than the algebraic sum, peaking at over L300 trillion from 1985 to 1986 (at 1989 prices) and not yet compensated by real gains in subsequent years. By the end of 1988, the loss for 1972-1988 inclusive still hovered at approximately L300 trillion. All conditions of real interest rates being equal, this would mean that household savers would recoup the losses of the 1970s and early 1980s in 1994 or 1995; *c)* in the 1970s, therefore, savers acted as lenders of last resort. After 1983, this role was assumed by the government budget. Since then, savers have passively looked on, in the hope of regaining previous losses.

TABLE 20

BREAKDOWN OF HOUSEHOLD FINANCIAL ASSETS

	1970	1971	1972	1973	1974	1975	1976	1977	1978	1979
Currency	9.9	9.8	10.1	9.4	9.6	9.3	8.8	8.3	7.6	7.0
Demand deposits	21.5	22.6	23.4	24.6	27.4	23.7	24.7	25.9	24.2	25.0
Other deposits:	34.9	36.9	36.4	36.6	42.4	49.0	49.9	50.1	47.1	45.3
- bank	22.6	23.4	22.5	22.8	28.9	34.8	36.2	37.0	34.5	33.0
- P.O.	9.2	9.8	10.2	10.3	10.0	10.8	10.8	10.5	10.0	10.1
- Spec. Credit CDs ..	3.1	3.6	3.7	3.5	3.5	3.4	2.8	2.6	2.6	2.3
- government agencies	0.0	0.0	0.0	0.0	0.0	0.0	0.0	0.0	0.0	0.0
Short-term sec.......	0.0	0.0	0.0	0.3	0.2	0.2	1.6	3.2	3.9	6.1
Bonds:										
- CCT										
- BTP	4.5	4.2	3.8	2.9	2.5	3.0	3.3	3.6	5.5	6.3
- Railway	1.2	1.1	1.0	0.8	0.6	0.5	0.4	0.3	0.2	0.2
- Other Spec. Credit	16.3	17.5	17.9	17.0	13.2	12.1	9.8	7.8	6.2	4.7
Shares.............	11.7	7.9	7.4	8.4	4.2	2.2	1.5	0.8	5.3	5.4
Total	100.0	100.0	100.0	100.0	100.0	100.0	100.0	100.0	100.0	100.0
TOTAL	51,439	58,375	68,351	82,575	90,649	109,484	130,003	157,262	201,902	244,237

Source: BANK OF ITALY: Annual Report, various years.

TABLE 20 continued

	1980	1981	1982	1983	1984	1985	1986	1987	1988	1989
Currency	6.3	6.2	5.8	5.4	4.9	4.5	4.1	4.0	3.6	3.8
Demand deposits	23.8	23.9	24.5	20.7	19.9	18.3	17.5	17.2	0.0	0.0
Other deposits:	40.3	37.4	38.2	35.9	33.8	30.9	28.2	28.1	42.2	41.1
- bank	29.1	27.1	27.3	25.9	23.9	21.6	19.0	18.3	31.9	30.5
- P.O.	9.1	8.2	7.7	7.1	6.7	6.6	6.7	7.2	7.1	7.3
- Spec. Credit CDs	2.0	2.1	3.2	2.9	3.1	2.7	2.6	2.6	3.3	3.3
- government agencies	0.0	0.0	0.0	0.0	0.0	0.0	0.0	0.0	0.0	0.0
Short-term sec.	10.2	14.8	14.7	15.3	15.8	14.9	13.0	14.7	15.5	16.6
Bonds										
- CCT		3.0	4.9	9.4	12.0	14.0	13.8	15.8	12.4	12.2
- BTP	4.9	2.6	1.8	2.1	2.4	2.0	3.5	4.5	7.3	7.9
- Railway	0.2	0.1	0.0	0.2	0.3	0.4	0.4	0.5	0.0	0.0
- Other Spec. Credit	3.6	3.2	3.3	3.6	3.4	3.1	3.0	3.3	3.4	3.6
Shares	10.7	8.8	6.6	7.3	7.4	11.8	16.5	11.8	15.6	14.9
Total	100.0	100.0	100.0	100.0	100.0	100.0	100.0	100.0	100.0	100.0
TOTAL	307,235	367,226	433,353	521,102	630,198	770,941	918,447	1,014,280	1,221,712	1,379,800

Source: BANK OF ITALY: *Annual Report*, various years.

NOMINAL AND REAL RATES OF RETURN OF SELECTED HOUSEHOLD FINANCIAL ASSETS
(in percentage)

TABLE 21

	(*)	1971	1972	1973	1974	1975	1976	1977	1978	1979	1980
Currency	n	0.00	0.00	0.00	0.00	0.00	0.00	0.00	0.00	0.00	0.00
	r	−4.76	−5.30	−9.42	−16.32	−14.68	−14.16	−15.33	−11.03	−12.89	−17.49
Demand deposits	n	3.50	3.50	3.75	5.91	5.87	7.85	8.51	7.71	7.50	8.27
	r	−1.43	−1.99	−6.02	−11.38	−9.67	−7.42	−8.12	−4.18	−6.36	−10.67
Other deposits:											
- bank	n	4.50	4.50	5.00	8.06	7.99	10.95	12.26	10.66	10.24	11.79
	r	−0.48	−1.04	−4.89	−9.57	−7.86	−4.76	−4.94	−1.55	−3.97	−7.76
- P.O.	n	2.50	2.50	2.50	3.75	3.75	4.75	4.75	4.75	4.75	4.75
	r	−2.38	−2.94	−7.16	−13.18	−11.48	−10.09	−11.30	−6.81	−8.75	−13.57
- Spec. Credit CDs	n	4.50	4.50	5.00	8.06	7.99	10.95	12.26	10.66	10.24	11.79
	r	−0.48	−1.04	−4.89	−9.57	−7.86	−4.76	−4.94	−1.55	−3.97	−7.76
- government agencies	n	2.50	2.50	2.50	3.75	3.75	4.75	4.75	4.75	4.75	4.75
	r	−2.38	−2.94	−7.16	−13.18	−11.48	−10.09	−11.30	−6.81	−8.75	−13.57
Short-term sec.	n	5.50	5.50	6.50	14.12	11.01	16.63	15.24	11.99	12.51	15.92
	r	0.48	−0.09	−3.53	−4.50	−5.28	0.11	−2.42	−0.36	−1.99	−4.36
Bonds:											
- CCT	n										
	r										
- BTP	n	7.04	6.58	6.85	9.18	10.25	12.61	14.71	13.18	13.12	15.30
	r	1.94	0.93	−3.22	−8.64	−5.93	−3.34	−2.87	0.69	−1.46	−4.87
- Railway	n	7.80	7.26	7.30	9.72	11.62	12.78	14.49	13.46	14.36	15.39
	r	2.67	1.57	−2.81	−8.18	−4.76	−3.19	−3.06	0.94	−0.38	−4.79
- Other Spec. Credit	n	8.02	7.29	7.29	9.64	11.11	13.13	14.61	13.52	13.78	15.65
	r	2.88	1.60	−2.82	−8.25	−5.20	−2.89	−2.96	1.00	−0.89	−4.58
Shares	n	4.36	3.40	2.50	2.95	4.89	4.15	4.58	4.91	3.18	2.43
	r	−0.61	−2.08	−7.16	−13.85	−10.50	−10.60	−11.45	−6.66	−10.12	−15.49
Consumer prices		5.0	5.6	10.4	19.5	17.2	16.5	18.1	12.4	14.8	21.2

(*) First six months of the years.
n = nominal; r = real.
Source: BANK OF ITALY: *Annual Report*, various years.

The Government Budget and the Italian etc.

TABLE 21 continued

	(*)	1981	1982	1983	1984	1985	1986	1987	1988	1989	1990(**)
Currency	n	0.00	0.00	0.00	0.00	0.00	0.00	0.00	0.00	0.00	0.00
	r	−15.11	−14.16	−12.82	−9.75	−8.42	−5.57	−4.49	−4.76	−5.93	−5.48
Demand deposits	n	9.32	9.89	9.50	8.84	8.21	7.04	6.19	6.10	6.32	6.32
	r	−7.20	−5.67	−4.54	−1.77	−0.91	−1.08	−1.42	−1.05	−0.02	−0.49
Other deposits:											
- bank	n	13.89	15.03	14.24	12.93	11.66	9.33	7.62	7.45	7.89	7.89
	r	−3.32	−1.26	−0.40	1.92	2.25	3.24	2.79	2.33	1.50	1.98
- P.O.	n	4.75	4.75	4.75	4.75	4.75	4.75	4.75	4.75	4.75	
	r	−11.08	−10.09	−8.67	−5.46	−4.08	−1.09	0.05	−0.24	−1.46	−0.99
- Spec. Credit CDs	n	13.89	15.03	14.24	12.93	11.66	9.33	7.62	7.45	7.89	7.89
	r	−3.32	−1.26	−0.40	1.92	2.25	3.24	2.79	2.33	1.50	1.98
- government agencies	n	4.75	4.75	4.75	4.75	4.75	4.75	4.75	4.75	4.75	4.75
	r	−11.08	−10.09	−8.67	−5.46	−4.08	−1.09	0.05	−0.24	−1.46	−0.99
Short-term sec.	n	19.70	19.44	17.89	15.37	13.71	11.40	10.73	11.12	12.58	12.58
	r	1.61	2.52	2.78	4.12	4.13	5.19	5.76	5.83	5.91	6.41
Bonds:											
- CCT	n	20.25	20.78	19.82	16.98	14.68	12.41	10.66	11.25	12.71	12.71
	r	2.08	3.67	4.46	5.58	5.02	6.15	5.69	5.95	6.03	6.53
- BTP	n	19.35	20.21	18.25	15.57	13.68	11.45	10.58	10.54	11.64	11.64
	r	1.32	3.18	3.10	4.31	4.10	5.24	5.62	5.28	5.02	5.52
- Railway	n	18.24	20.25	17.52	14.20	11.80	9.49	9.08	8.57	9.10	9.10
	r	0.37	3.22	2.46	3.07	2.38	3.39	4.18	3.40	2.63	3.12
- Other Spec. Credit	n	19.78	20.62	17.99	14.93	12.96	10.56	9.87	10.38	11.21	11.21
	r	1.68	3.54	2.87	3.73	3.44	4.40	4.94	5.12	4.62	5.11
Shares	n	1.89	2.24	2.45	3.09	2.69	1.58	2.01	2.57	2.50	2.50
	r	−13.51	−12.24	−10.68	−6.96	−5.96	−4.08	−2.57	−2.31	−3.57	−3.12
Consumer prices		17.8	16.5	14.7	10.8	9.2	5.9	4.7	5.0	6.3	5.8

(*) First six months of the year.
n = nominal; r = real.
Source: BANK OF ITALY: *Annual Report*, various years.

TABLE 22

GAINS (+) OR LOSSES (−)
IN HOUSEHOLD FINANCIAL WEALTH

	1971	1972	1973	1974	1975	1976	1977
Currency	−243.48	−302.43	−653.20	−1,267.42	−1,280.75	−1,435.71	−1,763.26
Demand deposits	−157.93	−262.60	−962.56	−2,307.74	−2,404.43	−1,928.91	−2,608.52
Other deposits	−175.46	−332.64	−1,376.03	−3,197.69	−3,345.02	−3,189.04	−4,103.20
Short-term sec.	0.06	−0.02	−0.92	−9.59	−9.93	0.24	−51.12
Bonds:							
- CCT	—	—	—	—	—	—	—
- BTP	45.33	22.65	−82.58	−210.03	−132.30	−111.12	−121.94
- Railway	16.96	9.92	−19.29	−53.69	−25.57	−16.86	−14.73
- Other Special Credit	240.48	163.77	−344.78	−1,161.17	−620.07	−382.47	−376.96
Shares	−36.58	−96.38	−360.08	−962.53	−395.66	−253.68	−224.26
Losses, at current prices	−310.62	−797.73	−3,799.44	−9,169.86	−8,213.73	−7,317.56	−9,263.99
Cumulative losses at current prices	−310.6	−1,108.3	−4,907.8	−14,077.6	−22,291.3	−29,608.9	−38,872.9
Cumulative losses at base year prices	−310.6	−1,125.7	−5,042.5	−15,195.4	−26,022.7	−37,634.0	−53,709.7

Source: BANK OF ITALY: *Annual Report*, various years.

TABLE 22 continued

	1978	1979	1980	1981	1982	1983	1984
Currency	−1,435.27	−1,966.54	−3,010.85	−2,939.57	−3,234.85	−3,240.02	−2,764.62
Demand deposits	−1,703.57	−3,106.41	−6,502.45	−5,269.19	−4,989.84	−4,818.41	−1,911.46
Other deposits	−2,090.08	−4,750.26	−10,026.8	−6,294.51	−4,407.03	−3,407.36	880.27
Short-term sec.	−18.21	−158.96	−646.93	504.76	1.367.79	1.777.14	3.296.83
Bonds:							
- CCT	—	—	—		409.67	953.11	2.719.03
- BTP	39.00	−163.17	−745.97	199.66	301.00	247.01	468.52
- Railway	4.09	−1.8	−24.11	−1.78	7.73	4.94	37.28
- Other Special Credit	122.96	−110.94	−526.65	183.61	415.04	415.25	692.78
Shares	−85.50	−1,074.95	−2,040.08	−4,441.43	−3,939.68	−3,055.77	−2,644.08
Losses, at current prices	−5,166.57	−11,333.0	−23,523.8	−18,054.9	−14,070.2	−11,124.1	774.55
Cumulative losses at current prices	−44,039.5	−55,372.5	−78,895.5	−96,950.4	−111,020	−122,144	−121,370
Cumulative losses at base year prices	−65,536.3	−86,568.7	−128,445.0	−169,363.0	−211,378.0	−253,574.7	−280,186.2

Source: BANK OF ITALY: Annual Report, various years.

TABLE 22 continued

	1985	1986	1987	1988	1989	1990
Currency	−2,625.71	−1,943.38	−1,677.19	−1,943.43	−2,636.10	−2,889.04
Demand deposits	−1,143.11	1,516.59	2,278.05	1,831.07	—	—
Other deposits	2,101.25	5,525.39	5,550.26	4,772.48	5,153.43	8,218.96
Short-term sec.	4,116.12	5,981,39	6,901.44	8,688.01	11,171.49	14,636.60
Bonds:						
- CCT	3,779.39	6,615.61	7,205.05	9,559.35	9,104.61	11,024.65
- BTP	632.41	824.95	1,804.66	2,422.98	4,492.28	5,989.04
- Railway	50.64	112.99	164.66	168.71	—	—
- Other Special Credit	745.60	1,060.01	1,340.45	1,706.69	1,910.97	2,515.80
Shares	−2,784.22	−3,720.01	−3,895.62	−2,763.35	−6,799.50	−6,412.85
Losses, at current prices	4,872.38	15,973.54	19,671.77	24,442.51	22,397.18	33,083.16
Cumulative losses at current prices	−116,497	−100,524	−80,853	−56,410.7	−34,013.5	−930.4
Cumulative losses at base year prices	−301,090.9	−302,881.8	−297,445.5	−309,875.2	−307,100.2	−291,723.2

Source: BANK OF ITALY: *Annual Report*, various years.

4. - Exploring the Causes of the Debt and Developing a Twofold Approach to Reform Italy's Public Sector: Who Should Pay the Price During the 1990s

As demonstrated in earlier sections, it seems possible to suggest a fundamental lack of coherence between the three targets of the public sector, to provide social services, state entrepreneurship and infrastructure, and the availability of resources. The stage of development reached by Italy in the late 1980s is not so much at issue as the lack of coordination between the setting of objectives and the accumulation of resources. The public debt is the embodiment of this disjointedness.

The current account deficit net of interest, the capital account deficit and public sector resources provided by means of endowment funds to state holdings and to ENEL are the flows which determine year after year the stratification of mounting debt stocks. It is common knowledge that the current account of the government budget presented surpluses until 1971, making it possible to partially finance the capital account and endowment funds. From 1971 on, the government budget consistently showed current account deficits. Debt soon began to spread, fueled by all three components.

In an attempt to determine the causes of the debt to date, we imagined what the public debt would have amounted to date if it had remained at the late 1969 level, rising only as a result of the burden of interest payments on the initial debt stock (Table 23). In 1989, this value would have equalled L274 trillion. Since the debt as at 31 December 1989 was equal to L1,168.3 trillion it can be deduced that the actual debt stock increase of L893.6 trillion was caused by the three "deficits" mentioned above and by the interest burdens generated by them. We then used the annual flows of the three deficits for 1970-1989 to reconstruct the debt stock created by each "source" (Table 24).

An initial "accounting" study of the table may lead one to conclude that the stock of public debt in 1989 was totally attributable to capital account deficits and to the funds paid to state holdings and to ENEL. As we can see, in fact, in Table 24, the debt which began to accumulate over the years following current account deficits (net of interest) and the burdens created as a consequence reached its peak in

TABLE 23

PERCENTAGE OF DEBT BROKEN DOWN
IN ACCOUNTING TERMS AS FOLLOWS

Year	Debt stock as at 31 Dec. (A)	Debt stock as at 31 Dec. 1969 yield bearing in subsequent years (B)	Debt accumulated over the years (C)=(A)−(B)
1970	37,636.0	36,430.5	1,205.5
1971	42,463.0	38,652.8	3,810.2
1972	48,762.0	40,856.0	7,906.0
1973	56,919.0	43,266.5	13,652.5
1974	65,756.0	47,030.6	18,725.4
1975	82,996.0	51,310.4	31,685.6
1976	102,396.0	57,159.8	45,236.2
1977	123,944.0	64,533.4	59,410.6
1978	158,206.0	71,761.2	86,444.8
1979	190,979.0	80,013.7	110,965.3
1980	228,554.0	90,895.6	137,658.4
1981	283,490.0	106,166.1	177,323.9
1982	362,007.0	124,214.3	237,792.7
1983	456,031.0	144,461.2	311,569.8
1984	561,489.0	164,541.3	396,947.7
1985	683,058.0	184,615.4	498,442.6
1986	795,597.0	203,261.5	592,335.5
1987	910,563.0	222,571.4	687,991.6
1988	1,035,263.0	244,605.9	790,657.1
1989	1,168,319.0	274,692.4	893,626.6

Source: BANK OF ITALY: *Annual Report,* various years; *Relazione generale sulla situazione economica del paese,* various years.

the creation of total debt in 1976 and in absolute values in 1983 Table 23. At the end of 1987, however, in accounting terms, the contribution of this debt is equal to zero.

This does not mean, however, that the "blame" for the debt cannot be placed in large part on current account deficits. Even with the necessary caution and verification, one has to consider that the capital account expenditure of the public sector, though deficit financed, can improve the conditions of development of industry, and

TABLE 24

DEBT ACCUMULATED AS A CONSEQUENCE OF DEFICITS IN:

Year	Capital account	Current account	Endowment funds	Current expenditure net of interest
1970	2,515.0	−1,042.0	247.9	−1,042.0
1971	5,073.4	−794.6	636.5	−2,269.0
1972	8,133.6	626.1	1,202.4	−3,672.5
1973	11,681.5	1,819.1	1,804.3	−5,384.4
1974	16,654.8	2,702.4	2,283.3	−7,706.3
1975	24,431.4	7,819.3	2,929.5	−10,543.6
1976	33,979.5	11,497.7	3,996.5	−14,538.3
1977	46,217.9	14,619.9	5,692.1	−19,937.0
1978	61,444.3	18,786.2	8,819.6	−26,500.0
1979	80,806.4	23,010.7	10,669.9	−34,762.7
1980	107,475.1	22,996.2	15,534.9	−46,378.5
1981	146,285.9	30,533.5	22,012.8	−62,703.0
1982	194,331.5	35,256.2	33,139.4	−83,893.9
1983	250,274.5	36,757.0	47,844.7	−110,435.4
1984	317,596.6	35,604.2	61,043.4	−140,452.4
1985	402,038.4	31,143.9	73,959.6	−174,225.9
1986	486,342.3	18,876.4	83,278.6	−210,658.6
1987	579,608.8	3,821.7	91,482.0	−251,350.1
1988	687,858.1	−17,391.9	100,795.3	−299,378.2
1989	826,296.7	−59,171.1	113,449.6	−362,573.9

Source: BANK OF ITALY: *Annual Report,* various years; *Relazione generale sulla situazione economica del paese,* various years.

by means of these improvements fuel the growth of current account revenue, which in accounting terms partially offset current account expenditure.

In section 2 we revealed contradictory factors at work: the overall structural effect of the government budget during the two decades appears to be negative in terms of productive capacity and full employment. However, a different composition of public sector expenditure with higher levels of investment expenditure seems to produce positive effects. It is therefore necessary to qualify the "accounting" aspects referred to above.

As outlined in Table 25, the Italian economy experienced a sharp rise in tax revenue, from 29% of GDP in 1970 to 41% of GDP in 1989. During the same period, current account expenditure net of interest rose from 27% to 38% of GDP. Line items therefore grew at approximately the same rate, revenue by 12 points and expenditure by 11 points. It is interesting to note that during the 1970s current account expenditure net of interest rose by 5 points of GDP from 1970 to 1975, while revenues remained steady at the same percentage of GDP until 1979. During the second five year period, while current account expenditure remained fixed at 32% of GDP, revenue rose by almost 3 points of GDP. Overall, during the 1970s, revenues grew less than expenditure net of interest and, in particular, they increased "five years later". Moreover, the 5 points of increase in current account expenditure net of interest realized during the first half of the decade was attributable to the "production subsidies" line item, for almost 2 points and to "social services" for more than 3 points. If we had to weight this figure, then we should indicate that during that period, two-thirds of the rise in current account expenditure were devoted to providing social services, and one-third was used to support industry.

During the 1980s, however, tax revenue expressed as a percentage of GDP increased 6 points during 1979-1983, higher by 1 point than the rise in current account expenditure net of interest. In recent years revenues continued to grow, reaching 40% of GDP in 1988, while current account expenditure net of interest remained steady at the 37% reached in 1983. Further, during 1980-1983, three of the five percentage points of increased expenditure are attributable to social services, while the role of "production subsidies" appears to be equal to zero.

Therefore, on the basis of these considerations, we sought to determine possible trends in the creation of public debt, if current account expenditure had increased at the same rate as current revenue. It is true that in this case as well, we are neglecting the possible effect of an increase in revenue caused by capital account expenditure. However, this enables us to estimate, albeit approximately, the amount of blame to be placed on the acceleration of current account expenditure, net of interest (having risen higher and earlier compared with the increase in revenue), in the creation of the public debt. The results of these calculations are outlined in Table 24. Given the assumptions mentioned, in 1988 the current account, net of

TABLE 25

CONSOLIDATED PROFIT AND LOSS STATEMENT
OF GENERAL GOVERNMENT:
CURRENT REVENUE AND EXPENDITURE,
NET OF INTEREST

Year	Revenue	Expenditure	Social services	Production subsidies
1970	29.0	27.4	11.6	1.4
1971	29.8	30.2	12.7	1.7
1972	29.8	31.7	13.5	1.7
1973	28.9	30.1	13.2	1.5
1974	28.3	28.9	12.8	1.5
1975	28.8	32.3	14.5	2.8
1976	29.8	31.4	14.4	2.5
1977	30.7	31.5	13.9	2.7
1978	31.9	32.9	14.8	2.8
1979	31.4	32.1	14.1	2.8
1980	33.2	32.4	14.1	2.9
1981	34.3	35.1	15.7	2.9
1982	43.3	43.2	19.5	3.7
1983	37.9	37.3	17.3	2.9
1984	37.7	36.8	16.7	3.1
1985	38.2	37.2	17.1	2.8
1986	39.2	37.5	17.3	3.1
1987	39.5	37.7	17.4	2.7
1988	39.9	37.9	17.5	2.5
1989	41.5	38.1	17.7	2.4

Source: ISTAT: *Conti delle amministrazioni pubbliche*, 1990.

interest payments, of the government budget had achieved surpluses capable of offsetting the growth of Italy's total public debt of over L363 trillion. The debt stock would then have equalled slightly more than L531 trillion as against the approximately L894 trillion actually realized. It should be noted that this result refers not to a block of current account expenditure expressed as a percentage of GDP but to its "growth" in quantity and over time compared with the trends in revenue which actually occurred. The provision of state social services and support to industry could have been achieved, if amounts and time periods had been coordinated during the 1970s, as the latest increase in social service expenditure was during the 1980s. As Giavazzi and Spaventa point out, it is true that the costs of the industrial restructuring of the 1980s are not the real causes of the current volume of public debt. Table 24 reveals that in 1988 the amount of debt accumulated as a consequence of capital account deficits equalled L688 trillion, while the amount of debt attributable to endowment fund grants to state holdings and ENEL stood at approximately L100 trillion. In accounting terms, the total of these two figures is equal to the total debt. On closer examination, however, this debt could have amounted to as little as L420 trillion, and the difference of L363 trillion is attributable to the premature acceleration of current account expenditure largely due, as demonstrated earlier, to the size of the increase and when it occurred.

Clearly, the government budget is faced with the necessity of a two-pronged approach to recovery. As in the case of any firm in similar difficulties, this strategy must make profit-and-loss statement and balance sheet adjustments.

The first objective requires a curtailment of expenditure on social services to suit current account revenue, while the second objective involves the necessity of privatizing a portion of government assets. To be sure, it is not so much a question of "putting assets on sale" as realizing that public sector assets are not as "public" as it would seem, since assets are overshadowed by a huge public debt. In reality, government assets belong to the holders of BOTs and CCTs. The government is in an awkward position, since it contracts indebtedness at much higher rates of interest than the rates of yield that it obtains on the assets listed on its balance sheet, and assumes all of the entreneurial risk involved by guaranteeing a certain yield to its debtors. However, this is purely from an entrepreneurial point of view: from a macroeconomic perspective, similar conclusions can be made if the real interest rate exceeds the rate of economic growth.

It should also be pointed out that the sale of a portion of government assets can only reduce liabilities and should not be confused with the profit-and-loss statement (the positive effect on the government deficit can be brought about by lower interest payment); this purchase should take place in the market and can only involve assets with a market value, provided that careful consideration is given to their strategic importance in the pursuit of public sector objectives and to the risk of excessive oligopolistic or monopolistic concentrations within the private sector.

One of the fundamental arguments of this paper has been the lack of coherence between the range of public sector objectives and available resources. Throughout our analysis we attempted to evaluate the measures necessary in a strategy for recovery. However, there is one last element deserving of mention: efficiency and effectiveness. The three objectives, the provision by the government of social services, state entrepreneurship and infrastructure, can be matched to existing resources by using them more efficiently. This requires the application of private sector, open market parameters to public sector activity. Logic dictates that priority be given to infrastructure, which increases the prospects of industrial growth, thereby generating resources to maintain social services. State entrepreneurship should follow, albeit "in the marketplace".

BIBLIOGRAPHY

[1] ALESINA A.: «The End of Large Public Debt», in GIAVAZZI F. - SPAVENTA L. (eds.): *High Public Debt: The Italian Experience*, Cambridge University Press, 1988.
[2] ALESINA A. - PRATI A. - TABELLINI G.: *Public Confidence and Debt Management: A Model and a Case Study of Italy*, NBER, 1989.
[3] BALDASSARRI M.: «Spesa pubblica e crescita economica», *Rivista di politica economica*, January 1983.
[4] BALDASSARRI M.: «Impatto del bilancio pubblico sulla allocazione delle risorse e sulla formazione di capacità produttive ed occupazionali», in VARIOUS AUTHORS: *Risorse per lo sviluppo*. Roma, Sipi 1985.
[5] BALDASSARRI M. et. AL.: «Un tentativo di formalizzare un approccio "Tobin-Modigliani" al risparmio», *Rivista di politica economica*, October 1990, forthcoming.
[6] BARRO R.J.: «Are Government Bonds Net Wealth?», *Journal of Political Economy*, vol. 87, November 1979.
[7] BARRO R.J.: «On the Determination of the Public Debt», *Journal of Political Economy*, vol. 87, November 1979.
[8] BERNARDI L.: «Dinamica di disavanzo delle entrate e delle uscite di contabilità nazionale: l'evidenza empirica dei grandi aggregati», Atti del seminario *Il disavanzo pubblico in Italia: i risultati di una ricerca*, Pavia, Collegio Ghislieri, 1983.
[9] BLANCHARD O. - DORNBUSH R. - BUITER W.: «Public Debt and Fiscal Responsibility» in CEPS: *Restoring Europe's Prosperity*, Cambridge (MA), MIT Press, 1986.
[10] BUITER W.: «A Guide to Public Sector Debt and Deficits», *Economic Policy*, vol. 1, November 1985.
[11] CAVAZZUTI F.: *Il nodo della finanza pubblica*, Milano, Feltrinelli, 1979.
[12] CIPOLLETTA I. - HEIMLER A.: «Processi di ristrutturazione, progresso tecnologico e crescita economica». *Rivista di politica economica*, no. 7-8, July-August 1989.
[13] CIVIDINI A., GALLI G. - MASERA R.S.: «Vincolo di bilancio e sostenibilità del debito» in BRUNI F. (ed.): *Debito pubblico e politica monetaria in Italia*, Roma, Sipi, 1987.
[14] COTULA F. - MASERA S. - MORCALDO G.: «Il bilancio del settore pubblico e gli effetti di spiazzamento: un esame dell'esperienza italiana», in COMIT, 1983.
[15] GIAVAZZI F. - SPAVENTA L.: «Italia: gli effetti reali dell'inflazione e della disinflazione», *Rivista di politica economica*, no. 7-8, July-August 1989.
[16] HAAVELMO T.: «Multiplier Effects of a Balanced Budget», *Econometrica*, vol. XIII, 1945.
[17] KEYNES J.M.: «The General Theory of Employment. Interest and Money», in *The Collected Writings of J.M. Keynes*, vol. VII, London, 1973.
[18] MAJOCCHI A.: «Il disavanzo pubblico in Italia, elementi per una valutazione critica», in GERELLI E. (ed.): *Il deficit pubblico: origini e problemi*, Milano, F. Angeli, 1984.
[19] MCCALLUM B.T.: «Are Bond-financed Deficits Inflationary? A Ricardian Analysis», *Journal of Political Economy*, vol. 92, 1984.
[20] MICOSSI S. - TRAÙ F.: «Finanziamento delle imprese e trasformazioni produttive: il ruolo della politica monetaria nel processo di ristrutturazione», *Rivista di politica economica*, no. 7-8, July-August 1989.

[21] MINISTERO DEL TESORO: «Towards a Public Debt Policy in Italy, Report by a Committee appointed by the Italian Treasury Minister», *Quarterly Review*, September 1989.
[22] MODIGLIANI F.: «Disavanzi pubblici, inflazione e le generazioni future», in MODIGLIANI F.: *Reddito, interesse e inflazione*, Torino, 1987.
[23] SPAVENTA L.: «The growth of public debt in Italy: Past experience, perspectives and policy problems», *Quarterly Review*, June 1984.
[24] SPAVENTA L.: «La crescita del debito pubblico: sostenibilità, regole fiscali e regole monetarie», in GRAZIANI A. (ed.): *La spirale del debito pubblico*, Bologna, il Mulino, 1988.
[25] TABELLINI G.: «Monetary and Fiscal Coordination with a High Public Debt», in GIAVAZZI F. - SPAVENTA L. (eds.): *High Public Debt: the Italian Experience*, Cambridge, Cambridge University Press, 1988.
[26] SYLOS LABINI P.: *Nuove tecnologie ed occupazione*, Bari, Laterza, 1989.

Index

advance-deposit scheme 86
Alesina, A. 9n
Andreatta, N. 20n
Associazione Promoteia (Bologna) 193–5, 197–8
balance of payments 125, 145, 177
Baldassarri, M. 8n, 9n, 181n, 192n, 197n
Banca d'Italia 13, 33, 38, 39, 115
  econometric model of 11, 112, 119–32
    price–wage sector 126–9
    results 132–53
banking system 14–15, 84–6, 100, 181, 189
  see also Banca d'Italia
Barbone, L. 91n
Barca, F. 20n, 38n, 62, 85n, 87n, 88n, 89n, 91n, 92n, 93n, 94n, 95n
Barro, R.J. 9n
Basevi, C. 125
Belgium 46
Bernardi, L. 8n
Berndt, E.R. 66n
Bianchi, B. 84n, 88n
Biscaini, A.M. 84n
Blanchard, O. 9n
Bodo, G. 34n
Briotti, G. 8n, 9n
Bruno, M. 24, 28
budget deficit 2, 8, 23, 29, 33, 43–4, 53, 115, 158, 193
Bundesbank 27
Butler, W. 9n

Calzolari, M. 125
Canovi, L. 94n
capacity utilisation 76–80, 114, 127, 128
capital account
  deficit 209, 211, 214
  expenditure 163–4, 169–70, 180, 210–12
Caranza, C. 91n, 92n, 98n
Cavazzutti, F. 8n, 194
Chan-Lee, J.H. 38n

Chile 192
Ciampi, C.A. 93n
Cipolletta, L. 20n, 61, 89n, 91n, 93n
Cividini, A. 9n
Colombo, C. 125
company debt 86–9, 91–2, 95–8, 103–4
competition 8, 14, 45, 71, 74, 78–9, 90, 115, 120–3, 127, 128–9, 143, 147, 173, 181
Confederation of Italian Industry (Confindustria) 62, 68n, 69n, 72, 73n, 79n, 81, 117, 197–8
Confindustria 62, 68n, 69n, 72, 73n, 79n, 81, 117, 197–8
Conti, V. 89n, 97n
Contini, B. 89n
Cottarelli, C. 98n
Cotula, F. 8n, 84n, 98n
credit 14, 84–5, 87, 89–90, 96, 98, 99, 115, 160, 173, 189–91
credit institutions 84–5, 99, 100
currency depreciation 23, 28, 71
current account
  balance 36–7, 90, 92–3, 123, 134–5, 141, 143, 144, 177–8, 199, 200
  deficit 12, 123, 159, 197, 209–11
  expenditure 163–4, 169–70, 211–14

D'adda, C. 20n, 38
Dal Co, M. 61
debt service 43, 47–8, 51, 158
De Cecco, M. 93n, 100n
deflation 11–12, 128
Delli Gatti, D. 96n
deverticalization 89, 96n
disinflation 20, 23, 37, 111–12
  1980–6 112–19
  Bank of Italy econometric model 119–32
    results 132–53
  exchange rate policy and 35–7, 112–13, 116, 120–9, 131, 135–6, 140–5, 149

218

Index

monetary policy and 112, 115, 120, 122–5, 131, 140–4, 149
  output cost of 39–46, 52–3
  role of interest rate 144–5, 149
dollar
  appreciation 44n, 93, 114, 125, 136–7, 150
  depreciation 11, 37, 87, 93, 112, 113, 125, 148, 150
  DM exchange rate 120–1, 125, 129–30, 136–7, 140–1
Dornbush, R. 9n, 42

economic growth 8, 13, 16–17, 21–2, 28, 29, 38, 39, 47, 49, 191–2, 214
  and technological progress 62, 63–9
  see also GDP
efficiency 16, 72, 95n, 96n, 181, 215
employment 19n, 20, 28, 38, 39, 40, 45, 81, 93, 95, 157–8, 192, 193–5, 196–200, 211
ENEL 209, 214
Engle, R.F. 120
eurodollar 154
European Economic Community (EEC) 27, 97n, 116, 128, 154
European Monetary Agreement 120, 142
European Monetary System 20, 23, 36, 37, 39, 41–3, 44, 52, 72–4, 87, 90, 93, 97n, 99, 111
European Monetary Union 14, 130n
exchange rate 10–12, 45, 51–3, 85–6, 96–8, 100
  and disinflation 35–7, 112–13, 116, 120–9, 131, 135–6, 140–5, 149
  depreciation 20, 24, 28, 29–39, 44, 86, 94–5, 201
  real 35–7, 44n, 57, 92–3, 101, 116
  role in inflation 11, 13, 28, 38, 42–3, 142–5
exports 24, 28, 29, 35–6, 38, 90, 95, 101, 133, 194

fair rent (*equo canone*) 119n
Fanna, A. 84n, 91n
Fazio, A. 85n, 91n, 92n
Federal Reserve 27
Filippi, E. 92n

fiscal drag 10, 33–4, 39, 45, 62, 87, 90n, 193, 201
fiscal policy 8, 10, 43–4, 50–2, 62, 85, 126
  see also specific policy
foreign currency 9, 86, 120, 122, 136
France 21, 22, 24, 26, 27, 38, 42, 64–5, 67, 70–3, 77–8
  government revenue and expenditure 161, 164, 168–9
  per capita income 166–7
  savings and financial balances of government 162–3, 168–9, 182–9
  taxes 165, 169–72
Frasca, F. 86n, 87n, 89n, 90n, 91n, 98n
Fua, G. 62
Fuss, M. 66n

Galli, G. 9n
GDP 8, 21, 33, 39, 40, 46–9, 53, 99, 134–5, 143–4, 158, 193, 194
Germany 21, 22, 24, 27, 38, 40, 52
  government revenue and expenditure 161, 164, 168–9
  inflation 20, 38, 39–40
  labour costs 26, 28
  per capita income 166–7
  savings and financial balances of government 162–3, 168–9, 182–9
  taxes 163, 165, 169–72
Giavazzi, F. 42, 44n, 46, 50n, 62, 86n, 87n, 90n, 91n, 94n, 213–14
Giovannini, A. 42n, 94n
Gordon, R.J. 144n
government budget 157–9
  causes of debt and reform of public sector 209–15
  lender of the last resort 1970s and 80s 200–8
  role of government budget in supporting industrial activity 191–200
  structure and workings of 159–60
    breakdown of savings and financial balances 177–89
    composition of government expenditure 173, 174–6
    expenditure, revenue and balances of 160–70, 213

government budget cont.

    industrial sector subsidies and taxes 173, 176
    tax revenue 171–6
    total domestic lending 189–91
    who receives expenditure 177, 178
    see also public expenditure
Graziani, A. 20n
Gressani, D. 43, 93n
Gros-Pietro, G.M. 88n
Guiso, L. 123, 147n
Gulf War 14

Haavelmo, T. 191
Heimler, A. 38n, 61, 66n, 88n, 93n
Hendry, D.F. 120
hyperinflation 31n
hysteresis 46
IMF 27, 28, 36
imports 86–7, 90, 120, 122, 123, 128–9, 133, 142, 145, 150, 154, 194
income distribution 87, 158–9, 200
income tax 29–30, 34, 52, 90n
incomes policy 115–17, 140, 145–7, 149
industrial relations 114–18
    see also trade unions
inflation 11–12, 20–3, 24, 26–8, 29–34, 36, 38–42, 44, 46, 49, 52–3, 56–7, 72, 76, 85, 87–9, 90n, 91, 92, 94, 96n, 118–19, 127, 128, 130–2, 201
    decline in 11, 111–15, 147–8
    effect of government budget on 192–5
    exchange rates and 13, 38, 93, 142–5
    internal and external components of 135–41
    monetary policy and 11, 99, 100, 122–3, 125
interest rates 11, 16, 43, 44, 47, 49, 51, 53, 85, 92–3, 102, 141
    nominal 49, 91, 115, 117, 122–4, 201, 204–5
    real 78–9, 87, 96, 99, 115, 117, 122–3, 131, 201, 204–5, 214
    role in disinflation 144–5, 149

investment 10, 20, 21, 28, 29, 38, 39, 45, 52–3, 76–80, 85, 88–9, 92, 95, 96, 99, 105, 126, 179, 189, 194, 196–7, 211
Ireland 42
Istat 94, 101, 102, 104, 129n, 175n, 176n, 213

Japan 7, 24
    government revenue and expenditure 160–1, 164, 168–9
    per capita income 166–7
    production 68–9, 71–3
    profit margins 77–8
    R. & D. 63–5
    savings and financial balances of government 162–3, 168–9, 180–9
    taxes 165, 169–72

Keynesianism 191
King, S.R. 144n

labour costs 24–8, 32, 34–7, 45, 72–6, 90–4, 101, 117, 127, 128
Landi, A. 86n, 94n, 97n
Larsen, F. 154
life cycle model 181n
Lira 44, 62, 85–6, 115, 130
    see also exchange rate
Llewellyn, J. 154
Lo Cascio, M. 15n
Lucas, R.F. 119

Macchiati, A. 91n
macroeconomic stabilization 20, 90–4
Magnani, M. 38n, 62, 85n, 87n, 88n, 89n, 91n, 92n, 93n, 94n, 95n, 105, 147n
Majocchi, A. 8n
Malavasi, R. 15n
Marotta, G. 86n, 87n, 89n, 90n, 91n, 98n
Masera, R.S. 8n, 9n
Meloni, E. 20n
Micossi, S. 86n, 90n, 92n, 99n
Milana, C. 38n, 61, 88n
Mitterand, President 42
Modigliani, F. 28n
Momigliano, F. 61

Index

monetary policy 9–12, 25, 36, 42, 43, 44, 52–3, 85–6, 111
   and disinflation 112, 115, 120, 122–5, 131, 140–4, 149
   and inflation 11, 99, 100, 122–3, 125
   and macroeconomic inbalances 90–4
   and restructuirng process 10–11, 94–8
   financial structure and 98–100
   see also specific policies
Monti, M. 85n
Morcaldo, G. 8n, 84n

OECD 39, 114, 154
oil 54, 131
   prices 11, 29, 111, 114, 125, 129, 130, 136, 141, 148, 154
   shocks 11, 20, 24, 52, 85, 88, 92, 93, 95, 111, 112–13, 122, 125, 129n, 131, 136, 143, 144, 150, 200
Onadu, M. 89n, 96n
Onida, F. 62
Onofri, P. 87n
output 20, 22, 24–5, 28, 35, 95
   and disinflation 23, 39–46, 52–3

Padoa-Schioppa, T. 28n
Pagano, M. 44n
Phillips Curve 127, 147
Piore, M. 62
Pontolillo, V. 84n
Prati, A. 9n
prices 11, 25, 38, 41–2, 46, 54–5, 69–72, 74, 77, 84, 94, 112, 128, 136, 139, 141, 146, 150–4, 191
   consumer 91, 111–14, 117–19, 122–3, 128, 131, 132, 134–5, 138, 144
   increases 30, 39, 56–7, 78, 87
   manufactured goods 113, 114, 122, 125, 143
   raw material 93, 114, 125
   relative 61, 115–16, 120–2, 125, 126, 143
   wholesale 36, 90–1, 101, 102, 111, 113, 128
private sector 12, 16–19, 42, 71, 78, 84–5, 90, 99, 172, 173, 189–91, 193, 196, 215

privatization 214
production adjustment 94–8
   economic policies and 84–90
productive capacity 180, 191–2, 196–200, 211
productivity 27, 34–5, 45, 74, 76, 87, 91, 93, 95, 96–7, 127
profits and profit margins 20, 21, 24, 25, 27–30, 34, 38, 44–6, 53, 76–80, 88, 89, 91, 94–5, 97, 102, 104, 114, 147, 148
PSBR 48, 189
public debt 8–10, 12–13, 14, 16, 22, 23, 43–4, 46–53, 99, 100, 157–9, 177n, 180, 197, 198–200
   causes of 12, 209–15
public expenditure 8, 11, 12, 14, 16, 48–9, 81, 86, 141, 157–8, 193, 196, 198–9
   see also government budget
public sector 12, 84, 114, 118, 126, 160, 172, 180, 189–92, 194, 198, 199–200
   reform of 209–15
   see also government budget

R. & D. 63–9, 81
Rebecchini, S. 86n, 90n
recession 9, 14, 20, 39, 41, 43, 114

resource allocation 61, 157–8, 191–2, 196–200
restructuring 13, 53, 61–3, 80–1, 100, 180, 199–200, 213–14
   adjustment paths in 94–8
   monetary policy and 10–11, 94–8
   production 8–10, 12, 69–72, 79–80, 93
   state subsidies for social security contributions and 72–6
Richard, J.F. 120
Riva, A. 86
Rosa, G. 88n
Rossi, S. 86n, 92n, 99n
Rubino, P. 119n

Sabel, C. 62
Sachs, J. 24, 40n, 42
Salituro, B. 87n
Saraceno, P. 96n

Saudi Arabia 114
savings 86, 162–3, 168–9, 177–89,
    192, 196–7
  household 177, 179–83, 186–91, 201–8
Scala Mobile system 115, 117–18, 127,
    128, 132
seigniorage 50–1, 52
Sembenelli, A. 94n, 97n
service sector 93, 114, 128, 148
Silva, F. 86
Silvani, M. 24
Sims, C.A. 119
Single European Market 100
Siracusano, F. 89n, 90n, 97n
social pact policy 115–17
social security 9, 29–30, 32–3, 62,
    72–6, 87, 91, 102, 117, 127, 163,
    165, 169, 172–6, 194, 196
social services 158–9, 169, 173, 209, 212,
    213, 214–15
social welfare reforms 10, 12, 49, 52, 200
Spaventa, L. 8n, 9n, 28, 46, 62, 86n, 87n,
    90n, 91n, 213–14
state holdings, endowment funds to
    209, 211, 214
State Participations Ministry 87
subsidies 29–39, 44–5, 155–7, 173–6, 212
supply shocks 20–1, 22, 24–39, 44–6, 53
Survey of Italy (OECD–1977) 27–8
Sutch, H. 38n

Tabellini, G. 9n
Tarantola, R. 84n
taxes 11, 29–39, 44–5, 49, 52, 70, 77n, 85,
    90n, 126, 141, 158, 163–6, 169–76,
    193, 194, 195, 196, 199, 200, 211, 212
technical change 61–2
technological progress 62, 76, 88
  and production costs 69–72, 79–80
  R. & D. and 63–9, 81
trade unions 32, 33, 34, 35, 45, 52, 61, 62,
    87, 117–18, 201
Trau, F. 89n, 90n, 96n, 97n, 98n
Tresoldi, C. 89n, 90n, 97n

unemployment 24, 40–1, 122, 128, 137n,
    138, 144

United Kingdom 21, 22, 24, 27, 36, 43–4
  disinflation 45, 53
  government revenue and expenditure
    161, 164, 168–9
  inflation 21, 23, 38, 39–40
  labour costs 24, 26, 28
  per capita income 166–7
  production 68, 71–3
  profit margins 77–8
  R. & D. 64–5
  savings and financial balances of
    government 162–3, 168–9, 181–9
  taxes 163, 165, 169–70
  unemployment 39, 40–1
United States of America 70, 93, 97n
  government revenue and expenditure
    160–1, 164, 168–9
  per capita income 166–7
  production 68–9, 71–3
  profit margins 77–8
  R. & D. 63–5
  savings and financial balances of
    government 162–3, 168–9, 181–9
  taxes 165, 169–72

Vaciago, G. 89n, 100n
Visco, I. 34n, 119n, 123, 127

wage supplementation fund 19n, 45,
    62, 93, 96n, 138n
wages 9, 20, 24–5, 28, 46, 58, 74,
    102, 114, 117, 173
  gross 29–33, 45
  indexation 24–5, 29–31, 33, 42–3,
    45, 52, 54–6, 87, 90, 118, 127
  nominal 24, 28, 34, 54–6, 126–8, 146
  real 23, 24, 29, 38, 39, 55, 90n,
    91, 93, 101
West Germany
  production 67, 70–3
  profit margins 77–8
  R. & D. 64–5
  see also Germany
Wyplosz, C. 40n, 42

Zacchia, C. 62
Zanetti, G. 88n, 89n